Powering Generations

The TransAlta Story, 1911–2011

Copyright © TransAlta Corporation, 2011
All rights reserved

Produced by: TransAlta Corporation, Calgary, Canada
Authors: Dr. Robert Page, David A. Finch
Project Management: Kingsley Publishing Services
Historian and Researcher: David A. Finch
Book Design: BookWorks

2011 / 1

Printed in Canada by Friesens

Canadian Cataloguing in Publication Data Available
ISBN: 9781926832067

www.transalta.com

Front cover
Aerial view of Horseshoe Hydro Plant, c. 1990s.
TRANSALTA COLLECTION

Back cover
Transmission lines at Keephills 3, 2011.
TRANSALTA COLLECTION

Inside front flap
Gifford Horspool at Spray Lakes.
TRANSALTA COLLECTION

Inside back flap
Summerview II wind farm, 2010.
TRANSALTA COLLECTION

DEDICATION

To all TransAlta employees and friends: past, present, and future.

ACKNOWLEDGEMENTS

It took the energy, knowledge, and dedication of many people to bring the TransAlta story to life. Retired and current TransAlta employees who have helped with this project include: Harry Schaefer, Walter Saponja, Marshall Williams, Dawn de Lima, Jennifer Pierce, Chuck Meagher, Maria Cuartas, and Elyse Nabata. We wish to thank all of you.

TransAlta operations as of August 2011.
TRANSALTA COLLECTION

CONTENTS

Introduction
1

CHAPTER ONE
The New World of Electricity
3

CHAPTER TWO
Turning on the Lights: The Early History of Electricity in Calgary
12

CHAPTER THREE
Built to Last: The Creation of Calgary Power, 1911–29
23

CHAPTER FOUR
Depression, War, and Survival, 1930–49
45

CHAPTER FIVE
The Move to Thermal Power, 1950–79
74

CHAPTER SIX
Recession and Market Turbulence, 1980–94
103

CHAPTER SEVEN
Preparing for the Future, 1995–2011
124

AFTERWORD
The Way Forward
149

Index
151

TransAlta Milestones
One Hundred Years and Still Growing Strong

1909
- By purchasing the assets of other companies and acquiring solid financial backing and engineering expertise, Calgary Power Company Ltd., based in Montreal, creates the foundation for an investor-owned utility in Alberta.
- Planning and construction begins for the Horseshoe Falls hydro plant on the Bow River west of Calgary.

1911
- Horseshoe Falls hydro project completed.
- Calgary's Mayor John Mitchell flips the switch on 21 May, delivering hydro-electric power from Calgary Power's lines to consumers. The *Calgary Herald* declares the event "one of the most important steps in Calgary's history."

1913
- Kananaskis Falls hydro project completed.

1915
- Calgary Power begins supplying electricity to customers in Cochrane—a move that foreshadows its rapid expansion to other communities in southern Alberta during the 1920s.

Mid 1920s
- Transmission system extended into southern Alberta as far as Lethbridge, and franchises are signed with fifteen Alberta towns.

1929
- Ghost Hydro project completed, adding enough new capacity to the system to allow the company to begin sending electricity to Edmonton in 1930.

1930
- Transmission system extended as far north as Edmonton.

1933
- Calgary Power begins installing electric motors in grain elevators. By 1937, more than four hundred elevators are running on hydro power.

1937
- Calgary Power tours its "Modern All-Electric Kitchen" around Alberta towns and villages, spreading the word about electricity. It is staffed by home economists, who demonstrate the latest appliances for the electric home, including refrigerators, coffee makers, washing machines, and food mixers.

Late 1930s
- Calgary Power scouts out new hydro power sites in the mountains, ready for the expansion that would arrive in the 1940s.

1939
- Calgary Power begins sponsoring the Calgary Stampede Parade to the tune of $5,000, or about $75,000 in 2011 currency. This marks the first official link between Calgary Power and the Calgary Stampede, another legacy institution in the city.

1942
- Calgary Power purchases the Cascade power plant from the federal government and upgrades the facility. It is the first of a series of new developments during World War II and in the boom that follows.

1947
- The Barrier hydro project on the Kananaskis River starts delivering electricity.
- Calgary Power Company Ltd. moves its head office from Montreal to Calgary and reorganizes itself as Calgary Power Ltd. The move underlines the company's roots in the Canadian west and its commitment to the community it serves.
- The discovery of crude oil at Leduc, Alberta, in February, sparks a second oil boom in Alberta. With its solid base, Calgary Power expands its infrastructure and, by 1952, is serving more than 1,600 oilfield customers.

1948
- Calgary Power creates Farm Electric Service Ltd., an innovative, not-for-profit subsidiary that works cooperatively with farmers to link their Rural Electrification Associations into the power grid.

1950
- Calgary Power has more than four hundred staff members in offices across Alberta.

- Six thousand new customers are added to the system, and twenty-four additional towns and villages receive power.

1951–55
- The last of the hydro expansion projects in the Bow River watershed comes online with the Three Sisters, Spray, and Rundle hydro plants in 1951, at Bearspaw in 1954, and Pocaterra in 1955. Hydro power continues to play a key role in the company's power production.

1956
- Though Calgary Power is producing 99 percent of Alberta's hydroelectric power in the 1950s, nearly all potential hydro sites have been developed. The company has chosen to invest in coal as the new fuel for its generating stations. The first unit at the Lake Wabamun thermal generating station, west of Edmonton, begins producing electricity, using natural gas initially.
- Calgary Power shares are offered for sale, first to employees, then to the general public.

1959
- Customer numbers have skyrocketed during the 1950s, up almost 300 percent by 1959. Calgary Power has expanded to almost every corner of the province. Farms served grow from 10,000 in 1950 to 37,000 in 1959. On average, the domestic consumer uses twice as much electricity by the end of the 1950s as compared to ten years earlier. The cost of living rises 25 percent during the 1950s, while the cost of energy falls 25 percent, making electricity the deal of the decade.

1961
- Calgary Power marks its first fifty years of operation.

1965
- Calgary Power embraces modern technology when it purchases its first computer to help process payroll.
- Calgary Power begins generating electricity at the Big Bend hydroelectric plant on the Brazeau River, using water from a jointly constructed multi-use dam in partnership with the government of Alberta.

1967
- Second unit commissioned at Brazeau.
- By the end of the 1960s, Calgary Power's customer base has increased by 30 percent. The cost of living continues to rise during the 1960s—up about 40 percent—while the price for Calgary Power electricity falls about 15 percent during the same period.

1970
- The first unit at the Sundance steam plant at Lake Wabamun begins generating electricity in 1970 for a growing province with a huge appetite for power.

1972
- Calgary Power completes construction of the Bighorn Dam on the North Saskatchewan River.

1973
- At the Sundance steam plant, an innovation is implemented to capture fly ash (a by-product of coal burning) using electrostatic precipitators. With 99.5 percent of the fly ash being removed from smoke stacks, the company leads the way in another environmentally important conservation measure.

1975
- Offering Alberta residents a chance to invest in their booming economy, Calgary Power launches "An Opportunity for Albertans," which results in citizens purchasing 1.3 million common shares in the company.

1977–78
- Sundance units four and five come online.

1979
- Calgary Power becomes an official member of the Northwest Power Pool, linking its electrical generating capacity with regional partners to create a more reliable and sustainable network.
- By the end of the 1970s, Calgary Power is generating about 70 percent of all power produced in Alberta. It serves customers as far north as Athabasca—the geographic centre of the province—as far south as the U.S. border, and the provincial boundaries with Saskatchewan and British Columbia.

1980
- Calgary Power increases the number of consumers it serves by 25 percent.
- Sundance unit six comes online.

1981
- Calgary Power changes its name to TransAlta Utilities Corporation to better reflect its province-wide operations.

1982
- TransAlta engages in a public participation process to negotiate the move of the hamlet of Keephills, near Wabamun Lake, which stands on land slated for a mine.

1984
- Keephills plant comes online.

1985
- TransAlta's central control facility moves from Seebe on the Bow River to the System Control Centre at headquarters in downtown Calgary.

1989
- By the end of the 1980s, TransAlta is producing 72 percent of the electricity in Alberta. About 94 percent of its power comes from coal, making it the largest coal mining company in the country, with 23 percent of Canadian production.

1990s
- TransAlta invests in facilities in Argentina, Australia, New Zealand, and other countries.

1993
- As TransAlta diversifies to take advantage of new markets and new fuel options, it builds natural gas-fired cogeneration plants in Ottawa and Mississauga, Ontario.

1996
- TransAlta wins one of the first environmental awards from the Climate Change Voluntary Challenge Program as it develops its sustainable-development initiatives.

1997
- TransAlta expands into energy marketing, eventually operating trading offices in Portland, Oregon, and Annapolis, Maryland, as well as Calgary.

1998
- TransAlta publishes its first sustainability report.

1999
- TransAlta purchases a coal mine and generating facility in Centralia, Washington, and, with its technical expertise, transforms the plant into one of the cleanest coal-fired facilities in North America.
- When the Alberta government begins deregulating the electrical industry in Alberta, TransAlta divests itself of its provincial retail and distribution businesses, choosing to focus solely on power generation.

2000
- TransAlta purchases Vision Quest Windelectric, a three-year-old utility in southern Alberta that owns and operates sixty-seven wind turbine power plants with 44 MW of total peak capacity.

2001
- TransAlta begins trading on the New York Stock Exchange on 31 July.

2003
- TransAlta consolidates its trading operations to its Calgary headquarters.

2009
- TransAlta purchases Canadian Hydro Developers and its wind and hydro-electric and biomass assets.

2010
- TransAlta purchases wind farms in Western Canada, Ontario, and the Maritimes.

2011
- By 2011, TransAlta has built or added fifteen wind power facilities to its operations.
- As it celebrates one hundred years of providing sustainable electricity to consumers and stable returns to investors, Canada's largest investor-owned utility continues to serve customers with innovative, reliable, and low-cost energy.

TransAlta CEOs

1909–10	**William Maxwell Aitken**
1910–11	**Herbert Samuel Holt**
1911–21	**Richard Bedford Bennett**
1921–24	**Victor Montague Drury**
1924–28	**Izaac Walton Killam**
1928–60	**Geoffrey Abbot Gaherty**
1960–65	**George Harry Thompson**
1965–74	**Albert Warren Howard**
1974–85	**Marshall MacKenzie Williams**
1985–95	**Kenneth F. McCready**
1996	**Walter Saponja**
1997–2011	**Stephen G. Snyder**

INTRODUCTION

Principled, Reliable, and Sustainable

For generations, TransAlta has been serving society, first in Alberta, and today across Canada, in the United States, and as far away as Australia. Delivering power to the customer has always come first—responsibly, sustainably, and with an eye to the future. While modern Alberta has become well known as the centre of the Canadian oil patch, the reliable provision of electricity has helped this western province and its people prosper since the early 1900s.

The story, and the company—initially called Calgary Power—started small. One hydroelectric facility at Horseshoe Dam on the Bow River began delivering electricity to the people of nearby Calgary in May 1911, a tradition that carries on to this day. Hydro was reliable, reasonably priced, and provided the power upon which a rapidly growing frontier city could build a future. Electricity from the Ghost Dam project came on stream in 1930, just as the economic downturn staggered the world, but Calgary Power's energy continued to support provincial growth even during the Depression.

The Ghost Dam facility more than doubled the company's generating capacity, and allowed it to build a high voltage power line from the Bow River plant to the provincial capital in Edmonton. Calgary Power's people rose to the challenge of the Depression. They rustled up new markets—villages, towns, and country grain elevators—and improved the company's infrastructure. By 1940, the company was delivering 42 percent more power than in 1931.

Calgary Power more than doubled its customer base during the 1940s, then quadrupled its growth in the 1950s. Farm Electric Service Limited, a subsidiary, expanded the company's service into rural areas, too, demonstrating a commitment to all Albertans. And Calgary Power ran transmission lines to thousands of oil wells and other petroleum facilities as Alberta began reaping the rewards of its hydrocarbon resources.

When the company had fully developed all viable hydro options on Alberta rivers, it responded from strength: it applied its best people, excellent technology, stable financing, and engineering expertise to the task of making electricity using thermal generation. Calgary Power purchased coal reserves and mining equipment in the 1950s, and expanded its facilities quickly during the post-war boom times. By 1980, its coal-fired plants were generating three times as much electricity as the hydro facilities, with more thermal plants under construction. And Calgary Power took early steps as a steward of the land, leading the way in land remediation at its coal mines, reclaiming the land in a sustainable manner before required to do so by law.

Delivering power to customers was always the first order of business, and Calgary Power delivered on that promise. In all weather and under all circumstances, the familiar Calgary Power service trucks were a common sight across the province. The company's people provided many other services to the communities where they lived and worked—from volunteer activities, to fundraising for the United Way, to growing gardens to feed the poor and elderly. These and many other stories fill this book, and help define the legacy of a company with deep roots in local society.

During the 1950s, the company also broadened its investor base. Upon the death of Izaac W. Killam in 1955—the company's fifth president and its owner—shares were offered to company employees and the public. But Killam's spirit lived on. "We make money," he told a young Marshall Williams, who eventually became president of the company, "but our purpose is to serve the customers. If you serve them well, the shareholders will do well."

Company leaders had to apply their considerable talents to a new set of challenges during the downturn of the 1980s. In 1981, in order to better reflect its growth, the company renamed itself TransAlta. It also broadened its mandate and pursued a growth strategy beyond Western Canada. Throughout this turbulent decade, TransAlta grew steadily, increasing its dividend and the amount of electricity it delivered to customers. It repelled a hostile takeover and found new markets. It conducted an early CO_2 test project in partnership with an oil company. TransAlta was on the cutting edge of innovation in this area, and the experiment proved that technology could scrub the greenhouse gas from thermal emissions and inject it into an oil field to increase productivity—a precursor to the carbon capture and storage technology being developed today at Project Pioneer.

It is said that success is reserved for those brave enough to experiment. As a result, TransAlta's story includes a few missteps, calculated forays into business ventures that did not provide a reasonable return on investment, and other learning experiences that made its leaders wiser. The company has also pursued international opportunities, not all of which have proven successful. But today's electrical generation operations in the United States and Australia are a legacy of these bold initiatives.

When the Alberta government began deregulating the utility industry in the 1990s, TransAlta made innovative changes to its structure in the face of these unprecedented challenges. Building on its Alberta roots, it solidified its international operations in the United States and Australia. It hired a leader from outside the corporation—a first for the company. It also narrowed its focus from a vertically integrated utility to become a streamlined power generation and energy trading company, and consciously pursued a broader fuel portfolio. TransAlta's legacy operations—hydro and coal—are now coming into balance with newer alternatives, including natural gas, wind power, and geothermal energy. As a result, in 2011 the company is Canada's largest publicly traded provider of renewable power. A full quarter of TransAlta's energy comes from renewable sources, and its people are constantly applying technology to lead the way into a future where energy is cleaner and more sustainable.

TransAlta's future promises to be every bit as fascinating as its past: who could have predicted the adventures of its first one hundred years? Its leaders have nurtured a company that is ready to meet the challenges of another century. TransAlta has never been a flash in the pan nor a speculative investment. It is a corporate citizen that fulfills the needs of the people it serves.

In 1986, TransAlta published *75 Years of Progress* to celebrate its anniversary. "Its continuing tradition of service, excellence and value has kept the company strong through good years and lean," the booklet explained, "enabling it to meet its goal of supplying safe, reliable electric service to Albertans at the lowest possible cost." The company has built on this tradition in the last quarter century.

Change has affected the company in many ways during the last twenty-five years. As it celebrates its centennial, TransAlta is a larger, more focused company. It remains dedicated to supporting sustainability in the communities where it operates, meeting customer demands for reliable energy, and providing good value to its investors. It stands by its legacy. TransAlta remains committed to delivering to the public the electricity that is so essential to quality of life, and to doing so in a responsible manner.

> "The idea of electricity binding the world together in a body of brotherhood is clear. Electricity offers the twilight zone between the world of the spirit and the world of matter."
>
> ELBERT HUBBARD, JOVIAN SOCIETY WELCOME ADDRESS, 1899

CHAPTER ONE

The New World of Electricity

A powerful new form of energy emerged in North America during the late 1800s. It transformed human activity as few events in history have done. It changed the way people lived, played, and worked. It eased the drudgery of manual labour for men in factories and women in the home; nothing would ever be quite the same. This corporate story opens with the birth and early evolution of electricity in North America, for it was out of these events that Canada's largest wholesale electricity generator was born. Today called TransAlta, the company's story is rich and enduring.

Electricity was important in its own right, but it became even more significant as an enabler of a thousand applications powered by its current. Each decade since the 1880s has seen ever-increasing uses for electricity. This steady growth in electricity consumption has been very positive for the industry, rewarding for shareholders, and life-altering for humanity.

The new electric lighting created a market for power sources. Part of the drive was true need and part of it was curiosity: what was this new energy form that turned night into day? Newspapers and magazines were filled with speculative articles, which only stimulated the public imagination. Today, it is hard to compare this excitement to any other event, except perhaps the first moon landing and space travel. The arrival of electricity triggered an entire new dimension for the human imagination, because no one was quite sure where it would lead.

Ferris wheel at Chicago World's Fair, 1893. CORBIS BE060177

This opening chapter explores the colourful beginnings of the electricity business in North America and sets the stage for the arrival of electricity in Canada in the late 1880s and Calgary in the early 1900s. It is a unique story in the history of North American business. It contains a lively cast of characters and remarkable events. It is, in many ways, similar to the early days of oil and gas development in Texas. However, electricity lacked the driving centralizing presence of a John D. Rockefeller to bring order out of the chaos. Dozens of small producers desperately attempted to generate enough revenue to survive. This erratic pattern was repeated in Calgary during the early days of electricity, before the arrival of financier and businessman Max Aitken and the birth of his new company, Calgary Power, which later became TransAlta.

Mystery and Invention

While the basic theory of electricity had been known for some time, no means had existed to generate and harness its power. A number of scientists, including Ben Franklin in the mid 1700s, had experimented to learn more. Electricity's most visible form was lightning, viewed with fear and awe because of its immense power. Some traditional societies had interpreted lightning to be an expression of the gods' displeasure.

Although electricity had obvious advantages, there were also risks and hazards. Wiring installations could be faulty, wires could short out, and fires could result, leaving property damage and personal injuries. Many early contractors had no professional training. They learned on the job and tried to cut corners, such as limiting the use of expensive copper wiring. Loose connections on outside wiring, with the resulting sparks, became a fireworks display for the locals. All of this brought a sense of excitement and uncertainty to the early installations.

There was also a sense of mystery about electricity. Most people had no idea how electrons moved through a wire or the filament of a

Fictional Electricity

Electricity captured the imagination of science fiction writers in the mid-1800s. So much had happened in a short time that writers were prepared to speculate wildly about the future.

In *Twenty Thousand Leagues Under the Sea* by Jules Verne—published in 1869—the fictional Canadian king of the harpooners, Ned Land, is captured along with narrator and scientist Pierre Aronnax and his manservant, Conseil. Captain Nemo shows them the *Nautilus*, an unusual ship powered entirely by electricity. "It is evidently a gigantic narwhal," the harpooner tells the scientist while chasing the fantastic submarine, "and an electric one." "Professor," said Captain Nemo, "my electricity is not everybody's. You know what sea-water is composed of ... chloride of sodium forms a large 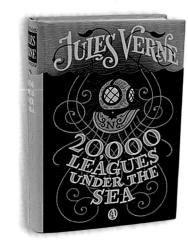 part of it. So it is this sodium that I extract from the sea-water, and of which I compose my ingredients. I owe all to the ocean; it produces electricity, and electricity gives heat, light, motion, and, in a word, life to the *Nautilus*."

Land, Aronnax, and Conseil eventually escape the *Nautilus* and the treacherous Captain Nemo, but with an exciting appreciation for the new world of electricity!

light bulb. Some linked electricity to the powers of the divine, while others viewed it as the embodiment of dark and sinister forces—which humans should avoid. These perceptions changed fundamentally in the 1870s with the development of dynamos to generate power, copper wire to move it, and the incandescent light bulb to employ it. The essentials required by entrepreneurs were now in place.

Electricity also captured the public imagination. It was seen to be in the vanguard of the forces of change in Victorian society. The new technology would usher in a new era of "progress" and prosperity for all. Humanity would be free of the need to work. There would be no separation between night and day, and no need for sleep. Women liberated from their domestic chores would be the intellectual equals of males, and electricity would lead to utopian social progress.

This new idealistic world would be led by the rational concepts of technology as interpreted by engineers and business leaders. Their success in the physical world would be followed by equal success in dealing with social issues or public affairs. It was almost as if society could be run like a gigantic electric grid. Across North America, engineering professionals and other like-minded technology advocates came together to form a group of societies that transformed the idea of technology into a type of theology. Elbert Hubbard, one of the leading spokesmen for this creed, put it this way: "The idea of electricity binding the world together in a body of brotherhood is clear. Electricity offers the twilight zone between the world of the spirit and the world of matter. Electricians are all proud of their business. They should be. God is the Great Electrician."

Thomas Edison

In the early history of electricity, Thomas Edison and George Westinghouse stand out for their leadership and their rivalry. Their efforts led to the formation of General Electric and Westinghouse, both corporate giants today. These companies were an expression of their founders' visions and became major equipment suppliers to every part of the continent, including Calgary. As a consequence of their sales efforts, these two firms also become owners or partial owners of individual utilities.

Of the two men, Edison was the more flamboyant. He was a curious mixture of scientist, businessman, showman, and larger-than-life American folk hero—the brilliant inventor who made good. Edison was an unabashed self-promoter with a folksy manner, which reporters admired. He loved to show visitors around his lab, explaining his many inventions and how they would change the world. His comments were often a combination of scientific fact and aspirational hope.

Edison's career was impressive. Born in 1847, he worked as a telegraph operator in Ontario and in several U.S. states. He experimented in both chemistry and electricity out of pure interest and he converted this interest into a career in 1876. His scientific genius was

George Westinghouse, c. 1900. CORBIS IH184468

Thomas Edison, c. 1878. EDISON NATIONAL HISTORIC SITE PHOTO ARCHIVES, 14.910/7

obvious, but his great challenge was raising the money to finance all his marvellous inventions.

While Edison worked on many projects with his colleagues, he found the greatest success with the generation of electricity and with products that used electricity. The advantages of Edison's incandescent light bulb were quickly evident to all. The light bulb was far superior to kerosene or whale oil lamps—the existing lighting—which gave off dirty, smelly fumes and darkened the walls and ceilings of Victorian rooms. The new lighting produced clearer, brighter light, gave off no fumes, consumed no oxygen, and was a lower fire risk. In addition to residential applications, there were commercial opportunities in street lighting, stores, theatres, and hotels.

Edison was the first to market the concept of district electrical generation and distribution. He recognized the economies of scale for district operations, as opposed to individual units serving each building. He also insisted on underground distribution, which was how gas companies delivered their product. Though safer, underground distribution was more expensive, and his rivals convinced city officials not to make it a requirement. This led to streets cluttered with the poles and wires of the various companies. As the number of rivals increased, Edison focused on supplying equipment to others rather than operating as an actual generator and distributor.

Edison launched his first community electrical system on Manhattan's Pearl Street, under the disapproving gaze of a few neighbours who feared he was installing an illegal whisky business. He tore up the sidewalk, installed a coal chute, and set up his boiler in the basement. Steam drove the dynamo that generated the power. The electricity was distributed through wires in tubes under the city streets to surrounding customers whose offices and homes had been wired by Edison. Even though the power plant was covered in coal dust and soot from the smoke, the inventor had it lit up like a Christmas tree as a publicity stunt. It attracted numerous visitors, some of whom became customers. Edison's purpose was to supply power to those in the area who wanted to use the new electric lights. Once he succeeded, entrepreneurs across North America sought to duplicate the business. The Age of Electricity had begun.

In a bid for market share, Edison and Westinghouse competed for the best type of transmission system: direct current (DC) or alternating current (AC). Edison chose the simpler, cheaper, and safer DC system, which, originally, made sense, especially for his systems. However, with time, it became less useful. The AC system allowed for power to be moved over longer distances. This became important, as power plants were often some distance from markets.

In 1890, North America slid into a steep economic recession and credit evaporated. Much of Edison's empire had been financed through debt in one form or another. As a result, Edison—always

Two Problems: Transmission and Storage

Two ongoing challenges facing electricity in its early years were transmission and storage. The first was costly, the second nearly impossible.

A number of early inventors experimented with ways to capture electricity from atmospheric sources. J. P. Morgan funded one such attempt in the 1890s when Serbian inventor Nikola Tesla tested the potential of suspended metal sheets to achieve electrostatic induction. The results did not meet expectations, and the concept was abandoned. Lack of storage has always been a major problem where electricity costs are concerned, since power is usually generated outside of the hours it was needed. The development of batteries helped, but they were large, heavy, and unconventional. Today we are still searching for a storage solution.

continued on page 9

6   POWERING GENERATIONS

Edison's Pearl Street plant in New York City, 1882.
CORBIS HQ001301

The Current: Then and Now

DC came first. Thomas Edison made direct current popular in the 1880s, and in 1887 there were 121 DC power plants in the United States. But there was a problem: after about a kilometre, DC electricity quickly lost power. AC, the alternating current system, invented by George Westinghouse, provided a solution. A high voltage AC system could transport power hundreds of miles with little loss. In 1895, Westinghouse built the first major power plant in North America at Niagara Falls and began moving electricity to distant customers.

Steam power was common in the 1880s, but it was awkward to use and could not be transported very far. When connected to a generator, steam power could be converted into alternating current, increased in intensity through a transformer into a high-voltage current, moved great distances, and then reduced by another transformer to levels suitable for use in a home or business.

Calgary Power relied on hydro power exclusively for decades, and later added thermal power—using natural gas and coal as fuels—to its system. TransAlta has added geothermal and wind power in recent years.

General Electric's Tower of Light at the Chicago World's Fair, 1893. CORBIS IH156312

short of capital—was particularly hard hit. He lost control of Edison Electric when he sold many of his shares to finance expansion, and General Electric achieved control.

Showcasing Electricity

The real turning point in popularizing electricity came with the Chicago World's Fair (1893) and the Buffalo Pan-American Exposition (1901). These exhibitions became great public showcases for electricity, with immense lighting displays. Westinghouse won the competition for the Chicago Fair, and the Buffalo Exposition showcased the new hydro power from Niagara Falls. People from all over the continent attended these huge events. Railways ran special excursions to the shows, and the public returned home with interest in all the new electrical equipment on display. In the public's imagination, electricity had become the way of the future.

After the Buffalo Exposition, there was greatly increased interest in hydroelectric generation as an alternative to coal. Many argued that the new "white coal" would trigger a second industrial revolution, because at the time water-power sites were cheap and plentiful, hydro required no fuel, and the rivers would run forever. Across North America a powerful movement supported hydro as the new clean energy solution. The Canadian federal government responded by sending teams to explore potential hydro sites, including Alberta's Bow River. If hydro was to be the energy source of the future, Canada wanted to become a leader in this new industrial revolution.

Challenges and Promises

The nature of the electricity business expanded during the early 1900s, as new applications for electricity emerged for the city, home, and factory. First came street lighting, then indoor lighting for stores, hotels, and homes. While arc lighting, because of its brightness, was the usual choice for the streets, incandescent light bulbs were the preferred option indoors. Lighting in stores extended the hours that shops could remain open, thus boosting sales. Electric lighting provided a warm and pleasant daylight environment for shoppers to view merchandise, and eliminating gas lighting improved air quality and decreased employee absenteeism. Business owners came to see this new electricity as key to attracting affluent middle-class buyers and as essential to commercial success.

Another important use for electricity came about with the development of the electric motor. Motors replaced much manual and animal-powered labour, increasing productivity and lowering costs. Water pumps completely altered municipal water and sewer systems. Electric fans improved ventilation in mines and factories as well as providing cooling and heating.

The invention of the electric motor was an historic moment, as depicted in this rather unusual collage from an unknown source. TRANSALTA COLLECTION

The Perfect Servant

At first, electricity frightened potential customers. It was not very dependable, and it was relatively expensive; most people relied on candles and lamps for lighting their homes. But, as electric lighting and appliances became standardized and more reliable, society began to accept the new technology. Soon the electric kitchen was all the rage. Motorized washing machines, vacuum cleaners, and other gadgets followed. Electricity became "the perfect servant."

Electricity also allowed cities to build higher and to develop infrastructure deep underground. Skyscrapers relied on electric elevators to move people high into the sky, and electricity provided light for the tall structures. Underground railways also used electric motors, and large cities developed subway systems to make transportation more efficient.

We live in an electric world today. Many homes rely on electricity for heat as well as power and are equipped with dozens of electric appliances. Transportation systems that use petroleum as a fuel are full of electrical components. Half the fuel consumed in North America becomes electricity. Lighting and telecommunications systems are completely electric.

Electric-powered elevators allowed for skyscrapers, while electric-powered streetcars enabled cities to expand into new suburbs, not to mention reducing the amount of horse manure on the roads. In factories, the electric motors revolutionized labour practices and increased output per worker.

New labour-saving devices also helped out in the home. Initially, the appliance market served only those who could afford the electricity system required to operate them. However, with North American mass production, prices came down and incomes went up. Mass advertising and the Eaton's and Sears home catalogues also drove the expansion. Electric stoves, washing machines, irons, and other appliances gradually became the norm for upwardly mobile middle-class families. The new manufacturing techniques introduced a consumer economy that demanded more electricity to power more electrical products. It was this market demand that led to the founding of companies like Calgary Power.

Electricity in Canada

Electricity came early to Canada because of its geographical proximity to the United States. Electrical engineer John J. Wright set up a trial electricity generator and supplied power to a few businesses in downtown Toronto in 1881, one year before Edison's district system was built in New York City. By 1883, street lighting systems were operating in Montreal and Winnipeg. In 1886, the Parliament Buildings in Ottawa installed their first lighting system, and by 1890 the streets of Ottawa, Hamilton, Pembroke, London, Victoria, Vancouver, Halifax, Saint John, St John's, Moncton, and Sherbrooke were alive with arc lights. The first electrical generating equipment arrived in Calgary in 1887, but it took until well into the 1900s before Calgary Power began providing reliable power.

Like Americans, Canadians were keenly interested in electricity's potential. Though it took decades for electrical appliances to become commonplace in most Canadian homes, Canadians were busy creating ways to use the new source of power. In 1874, Henry Woodward and Matthew Evans invented the electric light bulb—a patent Thomas Edison bought from them in 1879 and used in the manufacture of his incandescent light bulb. Joseph Wright came up with the idea for the electric streetcar in 1883. Thomas Ahearn invented the electric stove in 1882 and the electric car heater in 1890. Emile Berliner and Alexander Graham Bell both created phonographs in the late 1880s. Later on, Canadians invented the electron microscope, the oil-electric locomotive, the electric wheelchair, the electronic music synthesizer, an electric prosthetic hand, the artificial pacemaker, the electric organ, the television camera, the telephone, the wireless radio, the walkie-talkie, the wirephoto, and the agitating clothes washing machine. More recently, Canadians invented the Canadarm and the BlackBerry, as we continue to find new ways to put electricity to use in innovative ways.

From 1880 to 1914, electricity and its related technology evolved from mystery to necessity in the minds of many North Americans. Understanding the role of electricity explains much of the modern world that, today, we take for granted. Electricity brought lighting, streetcars, elevators, pumps, fans, telephones, movies, radio, and many other essential elements of our current economy and lifestyle. The new international corporate structures were only possible because of evolving communications and the new assembly line techniques introduced by Henry Ford. Electricity contributed to an era of innovation, efficiency, and entrepreneurship. It encouraged the belief that technology could solve every problem and create a perfect society—beliefs we know now were unsubstantiated. Above all, electricity became a symbol of the new modern age of light, heat, and movement.

The electricity industry matured and expanded beyond the frenetic entrepreneurship of the early days. The sector had developed into a three-tiered structure. There were equipment suppliers, like General Electric and Westinghouse, that dominated the field. There was an emerging group of centralized regional wholesale generators, such as the government-owned Ontario Hydro. And, at the local level, municipal utilities or city electrical departments distributed power to consumers. There was some overlap, however; for example, the General Electric holding company also owned shares in generating companies.

By 1914, many of these companies were well financed, had significant technical capability, and employed newly trained electrical engineering and trades graduates. It was a mature industrial sector with its own culture and language. Rapid expansion in the electricity market had created an important new industrial sector whose impact would eventually be felt around the globe.

> "When the bargain was completed all present expressed themselves well satisfied, and so will the taxpayers be pleased when the first of March comes and lights are turned on. Calgary will then not only be the best lighted but the cheapest lighted town in Canada." CALGARY HERALD, 26 JANUARY 1890

CHAPTER TWO

Turning On the Lights:
The Early History of Electricity in Calgary

When Thomas Edison launched his new power system in New York City in 1882, Calgary was a small, scattered group of houses on the prairies. Its history to that date gave little indication of its future importance. While First Nations had traversed the area for several thousand years in search of the massive buffalo herds, they did not have permanent settlements. Throughout the nineteenth century, Edmonton, with its Hudson's Bay Company trading post, was the centre of commerce, and it was the 1860s before Europeans began to appear in the Bow Valley. Some of these were American whisky traders who had trekked north from Fort Benton in Montana and had an armed camp near the present-day Glenmore Reservoir. These traders were a disruptive influence, triggering violence and social disintegration in their dealings with First Nations. The worst incident took place in 1873 about 300 kilometres southeast of Calgary. During the Cypress Hills Massacre, as it became known, American fur traders attacked and killed twenty-one First Nations people, including women and children, in revenge for the loss of horses the traders considered stolen.

These events attracted the concerned attention of the federal government in Ottawa. Sir John A. Macdonald, the Canadian prime minister, sent the North-West Mounted Police west to restore order and entrench Canadian sovereignty. He wanted to give Washington no excuse to send in the U.S. Cavalry. In 1875, the Mounties built Fort Calgary at the junction of the Bow and the Elbow Rivers.

Inside Horseshoe plant, c. 1947. TRANSALTA COLLECTION

Blood First Nations at Fort Calgary, 1878. GLENBOW ARCHIVES NA-354-23A

F Troop, NWMP, at Fort Calgary, 1876. GLENBOW ARCHIVES NA-354-10

Merchants, such as the Hudson's Bay Company and I. G. Baker from Montana, soon set up shop in the frontier town. The other prominent feature of the new settlement was the church. French-speaking Catholic missionaries who settled in what is today the Mission district gave Calgary a touch of Quebec. The signing of Treaty Seven in 1877 opened the way for white settlers by securing access to First Nations land in what would become southern Alberta.

In the late 1870s, cattlemen drove the first herds from Montana to the greener pastures north of the border. In 1881, the famous Cochrane Ranch was established west of Calgary, and other large ranches followed—with huge grazing leases granted by Ottawa. Calgary quickly became the supply centre for these ranchers, but movement of goods was difficult. The most effective transportation route was south, through the U.S.

Calgary's isolation changed suddenly and dramatically in 1883 with the arrival of the first train from the east. The Canadian Pacific Railway made Calgary its regional centre of operations and procurement. The railway needed coal, timber, food, and repair facilities. The community quickly became a bustling centre of activity. In 1884, Calgary formally incorporated as a town, with four hundred residents. To enhance its public image, it dropped the "fort" from its title, to try to separate its future from its unruly past.

Calgary was a CPR town. The company owned much of the land, employed most of the workers, and was the main purchaser of goods and services. But not all of Calgary's citizens were happy with the influence the CPR exerted.

The local political culture was a curious mixture that defied quick definition. There was the cowboy influence, which had migrated north from Montana. There were immigrants from Ontario. And there was a group of expats from England, including a few younger sons of nobility. The unruly individualism of the cowboy confronted the British reverence for law and order. It was a frontier, but a frontier of the British, not the American, Empire.

The new town council took up its duties with great enthusiasm

continued on page 16

CPR passenger train, Calgary, 1884.
GLENBOW ARCHIVES NA-967-12

The West's Iron Horse

The Mounties may have chosen the location for Fort Calgary at the confluence of the Bow and Elbow Rivers, but it was the steel track that gave life to the biggest city in southern Alberta. The first Canadian Pacific Railway train chugged across the Elbow River into Calgary on Saturday afternoon, 11 August 1883. And that iron horse delivered the printing press that produced the first edition of the *Calgary Herald Mining and Ranche Advocate and General Advertiser* three weeks later.

Until the train arrived, getting to Calgary took considerable time. The most reliable route was up from the south. Traders from Fort Benton in Montana started bringing supplies up the Mississippi and Missouri Rivers in 1846. Booze, too—so the Canadian government created the North-West Mounted Police in 1873 to patrol the west and control the whisky trade. To ship goods from the closest Canadian city, Winnipeg, was far more expensive and time-consuming than to go through Montana. Winnipeg was 1,300 kilometres away, compared with Fort Benton's 480. Long trains of bull-carts moved supplies from Montana to Calgary for several decades and operated for up to eight months of the year.

After the CPR arrived in 1883, Calgary boomed. It became the most important city between Winnipeg and Vancouver. In 1892, trains began running north out of Calgary to Strathcona, on the south side of the North Saskatchewan River from Edmonton. In 1898, tracks went south through the Crowsnest Pass into southeastern British Columbia. For the next half-century, the train was the king of transportation. Alberta grain and cattle went to continental markets by train. The CPR also developed irrigation in southern Alberta and even got into the oil and gas development game.

Round-up camp at A7 Ranch, Alberta, 1895. SASKATCHEWAN ARCHIVES BOARD, R-B1518

and not a few personal feuds. Local commercial rivalries were at the heart of many council debates, and there were no barriers to council members being financially involved in contracts under review. Concern over conflict of interest was for a later age.

One of the first actions of the new council was to create the Fire, Water, and Light Committee. The new town had to establish what constituted "essential" services for the community and whether the town should pay for them from the relatively small civic rate base. The only streetlights were a few murky coal oil lanterns, which on occasion provided target practice for drunken cowboys pouring out of the bars. "The question of street lighting has occupied the attention of the present Council on one or more occasions but nothing definite was decided upon," noted the *Calgary Herald* in an August 1887 editorial. "There can be no question as to the desirability of having the streets lighted, especially during the long winter nights and we think the Council would be justified in incurring the necessary expenditure for that purpose."

The first serious electrical supplier in Calgary emerged in 1885 when James Walker stepped forward. Using electricity from his sawmill, he installed a telephone line between his mill and his downtown store and later expanded the phone service to a few more businesses and linked them to a switchboard. In 1887, the powerful Bell Telephone Company arrived in Calgary from Montreal, and council began the search for a reliable power generator to supply Bell and to commence proper street lighting, which its citizens were now demanding.

Urban street lighting had both a practical and a symbolic role in early Calgary. It supposedly made the streets safer from crime and reduced injuries, since citizens could now see broken boards in the wooden sidewalks. Streetlights certainly improved the aesthetics and the pleasure of walking, especially on dark winter evenings. But lights were also about civic status. In 1894, with just over four thousand residents, Calgary was formally incorporated as a city—the first in what would be the province of Alberta. The new city claimed to be the most important urban centre between Winnipeg and Vancouver, and electric street lighting was part of the definition of a modern city. Streetlights, it was hoped, would help to attract new residents and businesses.

In 1887, the Calgary Electric Lighting Company emerged out of this scenario with a modest street lighting contract. The contract called for ten 40-watt streetlights at key locations for a monthly fee of $5 per light. Fortunately for the viability of the company, it had other customers. The debate over the smallest details of this contract saw the beginning of electricity politics in Calgary.

The Calgary Electric Lighting Company was a typical early Calgary enterprise. The company was capitalized at $25,000 through 250 shares at $100. The principals included Donald Davies and George King. Davies was a U.S. Civil War veteran, an I. G. Baker representative at Fort Macleod, and an MP in Ottawa. King was a former member of the Mounties, also an I. G. Baker representative in Calgary, and mayor from 1886 to 1888. Other shareholders included the

This photo, taken in 1953, shows one of the two remaining 1912 light standards that had been part of Calgary's first street-lighting system. GLENBOW ARCHIVES NA-5600-6163

Peter A. Prince

Born in 1836, near Trois-Rivieres, Quebec, Peter was the son of a Wisconsin millwright. He learned the trade in Eau Claire, Wisconsin, before moving to Western Canada in 1885. In 1886, he became the general manager of the Eau Claire & Bow River Lumber Co., a company with timber berths on the rivers west of Calgary. Prince worked for Isaac K. Kerr, an Ontario-born lumberman who was president of the Eau Claire firm in Calgary. When Kerr returned to Wisconsin in 1908, Prince continued running the lumber company. Importing technology from the United States, Prince and his son, John, built a steam-powered sawmill on the Bow River in Calgary. Soon it was making more than two million board feet of lumber each year for the booming city. By 1891, the Eau Claire lumber company was producing four million board feet of lumber annually. It ended its proud history after nearly seventy years of service to the community.

Peter Prince's original hydro plant, 1893. TRANSALTA COLLECTION

Peter Prince's sawmill on Prince's Island, 1895. TRANSALTA COLLECTION

Prince was a key person in early Calgary. He not only ran the lumber mill, but went on to create the Calgary Water Power Company in 1889. It gave the Calgary Electric Lighting Company, formed in 1887, a run for its money. The *Toronto Globe* noted: "The rivalry between the two companies gives the townspeople probably the cheapest electric lighting in Canada." Competition does that. Prince imported a 100 hp Corliss steam engine from Toronto and burned sawdust from the sawmill to provide electricity for the lumber company, the CPR facilities in Calgary, and some of the businesses and homes in central Calgary. By 1894, the company had the contract to supply electricity to the City of Calgary, a contract it held until 1905. His company eventually became part of Calgary Power in 1928.

Prince's Island Park in downtown Calgary honours Peter Prince, and the Eau Claire district on the banks of the Bow River takes its name from the Eau Claire & Bow River Lumber Co.

local manager of the Bank of Montreal and a prominent local doctor. None of the group appears to have had experience with electricity, and all had several other careers, including politics. They were typical local entrepreneurs who saw the need for the service and hoped to make a profit along the way. They brought limited capital and no technical expertise to their entrepreneurial role. It is not surprising that they adopted the inadequate but less expensive DC system.

Although the Calgary Electric Lighting Company signed a contract with the city in late 1889, streetlights were not turned on until March 1890. The company was suffering financial problems and went bankrupt in November 1892. And so, for two years Calgary streets were once again dark.

Calgary Electric's main competitor was Peter Prince, who ran the Eau Claire & Bow River Lumber Co. The company cut timber along the Kananaskis River and floated the logs down to Calgary, where they were sawed into lumber. Prince also linked the dam and sawmill on the Bow with a new hydro-power generation facility. This part of the enterprise was called the Calgary Water Power Company Limited, and it eventually won the city electricity contract and turned Calgary's lights back on in late October 1894. For the next decade, Prince enjoyed a monopoly.

Alberta Becomes a Province

Since 1875, the land that is known today as Alberta had been part of the North-West Territories. But after the 1904 federal election both Liberals and Conservatives agreed that the region should be granted provincial status. One province or two? The premier of the Territories favoured one large province with control over its natural resources. Prime Minister Sir Wilfrid Laurier thought two provinces of equal size would be easier to govern. He also kept control over natural resources as a way to encourage immigration to the new provinces. Alberta and Saskatchewan were born on 1 September 1905.

Red Deer considered itself the obvious choice for the Alberta capital, as it was halfway between Calgary and Edmonton. Banff, Vegreville, Blackfalds, and Athabasca Landing also offered themselves up to become the seat of government. But it was Edmonton that became the capital in 1906, much to Calgarians' displeasure.

Ranching was the first big industry in Alberta; the oldest ranches date back to the early 1880s. But the winter of 1906–07 almost spelled the end to the cattle business in the province. A harsh, cold winter, with few Chinook breaks to warm the air, resulted in the loss of tens of thousands of cattle and horses. The famous Bar U Ranch alone lost twelve thousand head. As a result, ranchers learned to put up hay against the winter threat.

The arrival of the Canadian Pacific Railway to Calgary in August 1883 provided a link back to central Canada. Wheat and other agricultural products were shipped out to central Canada on the train when crops were good. Homesteaders flocked west in large numbers after 1895, drawn by the offer of cheap land. The Calgary and Edmonton Railway linked Calgary to the capital in July 1891, and the Canadian Northern Railway arrived in Edmonton in November of 1905.

In 1914, oil was discovered at Turner Valley, southwest of Calgary, and thus began the quest for the black gold that has made Alberta rich. Subsequent discoveries of oil reservoirs at Leduc, southwest of Edmonton, in 1947, and the development of the oil sands at Fort McMurray in 1967, have allowed Alberta to become the wealthiest province in the country.

POPULATION OF CALGARY 1891–1941

1891	3,876
1901	4,398
1911	43,706
1921	63,305
1931	83,904
1941	88,904

Burns Block in boomtown Calgary, c. 1907. GLENBOW ARCHIVES NA-3752-6

The public debate that occurred about electricity service was more significant than the few dollars involved. It raised the issue of how this new energy source should be owned and managed. Some city residents felt strongly that street lighting was an essential service and therefore should be owned and run by the city; others considered it a commodity in trade that the city should monitor but leave to the private sector. There were similar debates about water services.

Another key policy issue was monopoly versus competition. It is generally accepted that competition in business is necessary to protect consumers. However, in the case of common carriers, there were strong economic arguments for monopoly, as a monopoly's large volume of business would provide the necessary scale for efficient delivery. In Calgary, as elsewhere, it took some time to sort out the issue. Eventually, in the first decade of the twentieth century, the City of Calgary's electric system emerged as the dominant player in distribution and retail. It gradually bought out local companies that lacked the capital and technology to serve city consumers. Where the private sector failed to meet market expectations, it could be replaced.

Edmonton followed a different model from Calgary for the generation of its electric power, reflecting a difference in political culture. The northern city purchased the Edmonton Electric Lighting and Power Company in 1902, a utility founded in 1891 by local entrepreneurs. The utility was both a generator—with the thermal Rossdale power plant—as well as a distributor. Edmonton's acquisition of the lighting and power company made it one of the first cities in Canada to own its own electrical utility. This model became an Alberta-based example for those in Calgary who preferred city-owned power generation. Calgary Power eventually extended its transmission lines north into central Alberta, surrounding Edmonton and serving communities far into northern Alberta.

A third option for electricity generation and distribution emerged with the development of the provincially owned Ontario Hydro. There the province created central generation and owned the interurban transmission to supply municipal utilities. Many Ontario cities wanted access to Niagara Falls hydroelectric power to help them attract industrial development to their communities. They feared that private sector Toronto interests would soon control the resource. This "public power" movement had supporters right across the country, including Alberta.

Peter Prince had solid backing from Wisconsin lumber interests who anticipated strong economic growth in Alberta. He wisely adopted the more mobile AC power system, as he was some distance from town. Prince was the first successful electrical entrepreneur in Calgary. He capitalized the company at $100,000—a significant sum for local companies of the day. He carefully managed the political, technical, and financial aspects of the job and secured Senator James Lougheed and lawyer R. B. Bennett to do his political and legal negotiations.

But hydro generation turned out to be more difficult than Prince had anticipated. Ice jammed and damaged the equipment in winter, and the seasonal variations in flow cut the generating potential. In frustration he fell back on his wood-fired boiler. He made improvements as he went, and Calgary Water Power Company remained profitable.

Calgary's slow rate of economic and population growth, which had plagued the city in the 1880s and 1890s, changed suddenly with the arrival of the new century. Boom times in the United States, the United Kingdom, and Western Europe had created great demand for Alberta products, and immigration flowed into the province as never before. The new refrigerated beef railcars and ocean steamers were a boost to the cattle trade. Investment poured in and business boomed. There were a few setbacks in the first decade, such as the crippling winter of 1906–07, but the city still experienced the largest expansion in its history.

Calgary's first electric streetcar, 1909.
GLENBOW ARCHIVES NA-644-22

The booming city strained the existing electrical generation system. The CPR had built branch lines out in every direction to service the influx of settlers, and a main line north to supply the Edmonton market. Its rail yards at Ogden in east Calgary were a major component of the local economy. The cattle stockyards were overflowing. Based in Calgary, Pat Burns created the greatest meat-packing empire in Canadian history. The city became an important centre, supplying settlers, farmers, and ranchers. Implements, building supplies, home furnishings, and new electric appliances filled the stores. For those unable to shop in the city, Eaton's offered house kits for homesteaders, delivered to the nearest railway siding. Electric streetcars began servicing Calgary suburbs in 1909, and an incredible air of optimism and confidence filled the city.

Although Calgary Water Power saw its profits increase during the boom, Prince followed a very conservative approach to further investment. There was no strong pressure from competitors and, as a result, the company did not focus on expansion. And because its profits were good, Prince chose to avoid risks and plough back high returns to shareholders rather than invest in new power sources. City council, needing more electrical capacity to sustain the high rate of economic growth, grew frustrated.

As a result, the city built a coherent electricity strategy through its own electrical department. It opened its own generating plant on Atlantic Avenue, fired by coal from Bankhead, near Banff. Peter Prince—by then nearly seventy years old—was no longer the dynamic entrepreneur he had once been and failed to challenge the new competition. The city took over the exclusive operation of street lighting, but resisted creating a publicly owned electricity monopoly. It was this ambivalence that allowed the private sector to survive.

After World War I, Calgary Water Power Company struggled on with steady profits but a declining share of the electrical market. The city's electrical department serviced all the new suburbs, and though the city made overtures to buy out Prince, a deal could not be secured. Peter Prince died in 1925. Prince's original idea to use hydropower was sound, but he had the wrong site and neither the capital nor expertise to dam an upstream site with any real potential.

Calgary's electricity industry struggled under an immature structure from the 1890s until World War I as it evolved from a frontier settlement to a town with business links to all the major North American centres. In these early days, private entrepreneurs and the city's electrical department worked together to supply service to a growing community. Meanwhile, there were fundamental changes taking place in the electrical utilities systems in central Canada and the United States. Technologically, small-scale, coal-fired generating facilities were being replaced with larger regional plants throughout North America, and many of them were harnessing the abundant potential of large river systems. Calgary's coal-fired facilities and limited water-wheel generation capacity simply could not keep up with the demand for electricity during boom-time growth.

Only one local group, the Calgary Power and Transmission Company Limited, was even attempting to develop new hydro sites on the Bow River. Its owners had correctly estimated the hydro potential of the river and secured a marketing contract with the City of Calgary. But they lacked the financing and the technological skills to tame the turbulent river. A new approach was needed to reduce costs, improve reliability and efficiency, and develop the economies of scale that would support a utility with the capacity to serve Calgary and, later, perhaps, the province. Into this challenging situation stepped a new group of businessmen, financiers, and visionaries who would fashion a utility company large and confident enough to provide electricity for the new West.

"Sunday morning at 8:30 Mayor Mitchell put in the plug which turned the hydro-electric power from the Calgary Power Company's lines over the city wire ... it marks one of the most important steps in Calgary's history." CALGARY HERALD, 22 MAY 1911

CHAPTER THREE

Built to Last:
The Creation of Calgary Power

Calgary was an exciting place in the early 1900s and witnessed the biggest boom of its history during the first decade of the twentieth century. New office and hotel buildings were being built downtown, while a new ring of suburbs extended the city out in every direction. Electric streetcars moved passengers, and electric commercial signs lit up the night. Calgarians had money in their pockets and a sense of confidence in themselves and in the future.

Every train brought new settlers from central Canada, the United States, or Europe. The open lands of southern Alberta filled up with ranches and homesteads. Grain and cattle prices were high, and the demand for labour exceeded the supply. Calgary was emerging as a major centre in the Canadian economy.

The economic boom stretched power supplies to their limit, triggering public concern. On 29 March 1910, the Calgary *Morning Albertan* published a strongly worded editorial headlined "Calgary and Cheap Power," demanding more electricity.

The city must either come to terms with the company at once or get additional equipment. The city is almost up to the limit now, and unless something is done about it will be past the limit by next winter. If the city is to maintain its power and light business it must increase the equipment one way or the other.

Calgary needed more power for street lighting, for streetcar

Kananaskis Dam, c. 1947. TRANSALTA COLLECTION

expansion, and for servicing new suburbs, such as Elbow Park. Both Calgary newspapers called for immediate action by city council to purchase Peter Prince's Calgary Water Power Company and to construct a new coal-fired power plant in the city to avoid power shortages in the coming winter months.

Although the Calgary economy was booming, its political climate had suffered a few shocks. When Alberta became a province in 1905, the Ottawa Liberals rewarded their friends in Edmonton by making the northern city the provisional capital. This reverse for Calgary was compounded when Edmonton became home to the University of Alberta in 1908—an honour that had been promised Calgary when it lost out in the bid to become the provincial capital.

In the first provincial election, the Edmonton-based Liberals trounced the Calgary-based Conservatives. The two party leaders were a study in contrasts. Tory leader R. B. Bennett from Calgary was a polished speaker and vigorous debater, while Alexander Rutherford was never comfortable in public debate. Rutherford quickly established a strong patronage system, which only further alienated Calgary. The Tories, in turn, had trouble mobilizing rural support, and Bennett became disillusioned with the frustrations of provincial politics.

Bennett was a curiously distinctive figure in frontier Calgary. Unmarried, he was a church and temperance leader in a community where drinking was an integral part of the local culture. His dress code also set him apart. His high-winged collar and tie with dark business suit stood out on the dusty streets of Calgary, where most dressed less formally. His business activities were also a matter of local comment. With his law partner, Senator James Lougheed, he defended the interests of the CPR and central Canadian banks in court, which was not popular with many local folks who detested these institutions. On the other hand, if you wanted deals completed, "RB" was your man.

With the economy booming, the provincial Liberals embarked on a series of government projects. They created the successful Alberta Government Telephone in 1907 as an alternative to the Montreal-based Bell Telephone. There was talk they would do the same for electricity, but that did not materialize. If it had, Calgary Power would have been terminated just as it was being launched.

But the Liberal government made one great mistake. This was an era of massive railway construction, before the full consequences of the arrival of the motorcar were understood. Two federally backed transcontinental railway systems were due to be constructed through Edmonton. However, the premier was concerned that the northern hinterland of the province was being ignored. The provincial legislature guaranteed the Alberta and Great Waterways Railway to the tune of $7.4 million in 1909—or the equivalent of $150 million in

Alberta Legislature building under construction, Edmonton, 1911.
GLENBOW ARCHIVES NA-1042-8

continued on page 26

R. B. Bennett: "Servant of All"

"I will end unemployment or perish in the attempt!" Such was the bravado of the flashy Calgary lawyer named Richard Bedford Bennett in 1930. That year he made $262,176—or about $3.2 million in 2011 currency—most of it in dividends. In the end, the economic Depression of the 1930s almost brought the proud westerner to his knees.

Not that Bennett had always been rich. Born the eldest of six children in 1870 in Hopewell Hill, New Brunswick, he was the son of a man of modest means. "I'll always remember the pit from which I was [dug] and the long uphill road I had to travel. I'll never forget one step," Bennett said many years later. He trained as a teacher and saved money to attend law school at Halifax's Dalhousie University. He arrived just like everyone else in early Calgary, to dirt streets. When he got off the train in January 1897 a still wind and -40 C temperatures welcomed him. He had to lug his own bags to the Alberta Hotel.

As a young man, he had responded to an invitation from Senator James Alexander Lougheed to serve as a junior lawyer in his Calgary firm. Arriving just as the humble cowtown began to grow, Bennett got in on the real estate boom, and was an investor in the Calgary Petroleum Products Company discovery well at Turner Valley in 1914—the well that put Alberta on the oil maps. Bennett first entered politics in the west in 1898 as the member for West Calgary in the government of the North-West Territories. In 1909, he won a seat in the Legislative Assembly in Alberta, and in 1911 the people of Calgary sent him to the House of Commons.

That same year, he accepted the presidency of Calgary Power, a position he retained until 1921 and in which he worked closely with his long-time friend, Max Aitken. Bennett's interests in numerous companies kept him busy and afforded him considerable income. In the 1920s, he was a director in E. B. Eddy, as well as Alberta Pacific Grain, of which he was president. Dividends from these companies as well as Canada Cement added to his wealth. He was also a director of Metropolitan Life Insurance of New York and was on the board of the Royal Bank of Canada.

R. B. Bennett, c. 1917.
TRANSALTA COLLECTION

In 1925, he once again ran for the House of Commons and won the seat for Calgary West. He became leader of the Conservative party in 1927, and though he had grown rich from working hard he promised to resign his company directorships to avoid any conflicts of interest. He also promised to be the "servant of all," and soon the chance was provided for him to live up to his word: he was elected prime minister of Canada. This first prime minister from Alberta never smoked nor drank, always dressed formally, and worked hard. He applied his lifelong discipline to the challenges of the 1930s. But the worldwide economic downturn was more than one man could conquer. As prime minister until 1935, he implemented numerous unsuccessful government programs to pull the country out of the economic depression. He favoured old age pensions and unemployment insurance. The Canadian Wheat Board and the Canadian Radio Broadcasting Commission—the predecessor to today's CBC—were created during his time as leader of the country. Letters arrived every day from Canadians, asking for help, and he said they "make life almost unbearable." He responded generously, giving away an estimated $2.3 million from his own pockets between 1927 and 1937—more than $30 million in 2011 equivalent.

Bennett retired from politics in 1938 and moved to England, where he was elevated to the House of Lords. He was also the Honorary Colonel of the Calgary Highlanders from 1921 to 1947, the year he died.

Upon Bennett's retirement from politics, economics professor Harold Adams Innis wrote: "Your leadership of the party especially during the years when you were Prime Minister was marked by a distinction which has not been surpassed ... No one has ever been asked to carry the burdens of unprecedented depression such as you assumed and no one could have shouldered them with such ability. I am confident that we shall look to those years as landmarks in Canadian history because of your energy and direction."

Laying the Alberta and Great Waterways Railway track, Bon Accord, Alberta, 1914. GLENBOW ARCHIVES NA-1649-20

2011 currency. The project proved to be a financial disaster. High costs per mile, poor management, excessive fees to central Canadian companies, and money that could not be traced all added up to a scandal that ended the premier's career in 1910. Bennett loved every last detail. For years he used the railway scandal to remind Alberta voters of the inevitable results of government ownership and provincial "socialism."

Premier Alexander Rutherford, c.1909. GLENBOW ARCHIVES NA-1514-5

The new Liberal premier, Arthur Sifton, was more able and less partisan than his predecessor, and more conciliatory to Calgary interests. The intensity of the partisan feuding declined when Bennett moved to federal politics with the election of 1911. Bennett's exit from provincial politics allowed him to become the president of Calgary Power from 1911 to 1921, and his political connections would prove vital to the young company.

Hydroelectric Power: The Vision and the Need

It is hard to imagine the sense of excitement and optimism that hydroelectric power generated in the early 1900s. Advocates claimed that this "clean" energy source would trigger a new industrial revolution. The old coal-based utilities would be swept aside by the new hydro, inexpensive and plentiful. Cleaner cities, healthier workers, and a higher quality of life would follow. Electric motors would revolutionize labour and industrial practices, and low-cost fuels would result in better wages. Countries with hydro potential, including Canada, would quickly surpass societies relying on coal-based power.

For Calgary citizens, there was a sharp visual contrast between the clean image of the new hydro and the existing power plants in the city. The latter were small-scale, polluting, and high-cost power sites covered with coal dust—a striking contrast to the images of Niagara Falls that the hydro proponents circulated. The Ottawa-based Commission of Conservation, founded in 1908, had explored the hydro potential of the Bow River, promoted it through reports, and discussed it with potential developers. The federal government also wished to promote efficiency and resource conservation and, through its management of Banff National Park, controlled development of the headwaters of the Bow River, west of Calgary.

The Bow River possessed many of the physical attributes that hydro advocates deemed necessary. However, local contractors were reluctant to tackle the engineering challenges of dams and power plants as well as the unpredictable force and fury of the Bow in flood. There were also worker safety issues to consider when building dams in the bed of a turbulent river and moving equipment to remote areas.

The main problem was the extreme seasonal variability in the water flow. The Bow River runs through dry lands on the eastern side of the Rockies, but its headwaters rise in the high country. Meltwater from Bow Glacier births the river, with mountain snow packs melting in the spring to provide most of its flow. By late summer the river is once again low, and in winter the Bow flows at a fraction of its June flood levels. Winter precipitation is snow, which sits for months before joining the river. Ground water and underground springs also contribute to the river, but are limited in volume. This seasonal variability in the flow of river water was not fully understood by the first studies and led to unduly optimistic estimates of the power potential and the profitability of hydro sites on the Bow.

Tough engineering challenges were also underestimated. The underlying geology of the riverbed was critical to the structural foundation of proposed dams. There were concerns about fractures or cracks in the underlying rock that might allow leakage under the dam. The width of the valley was important in estimating the generation potential. A wide valley added significantly to the size and the costs of a structure, but it also increased a reservoir's storage capacity and allowed delivery of electricity when the market price was highest. In the electricity business it's well known that you cannot store electricity: you can only store water.

Another major consideration when siting hydro power is the height of the water drop, or head. The higher the head, the greater the generating potential. In some cases a power canal down a valley could increase the head without significant additional cost. As heavy equipment was not yet available, dam construction projects had to rely on manual labour—using picks and shovels and heavy draft horses. Any business plan for a hydroelectric project had to include all these factors as well as the transmission costs of getting the power to market. Such estimates required careful assessment by seasoned professionals.

These factors help explain the dearth of Alberta contractors bidding on hydro contracts during this period. Any prospective contractor had to have extensive hydro engineering experience, deep pockets to cover cost overruns, political contacts in Ottawa to ensure the needed approvals, and good relations with Calgary City Council in order to market the power.

Beginnings

By chance, the wizard of Canadian finance, Montreal's Max Aitken, was visiting Western Canada in the early 1900s. Travelling between Calgary and Vancouver on the CPR, the man who was later bestowed the title of Lord Beaverbrook spied potential hydro sites along the Bow River. "As I was passing through the Kananaskis District, I saw what looked like a very fine power proposition within a hundred yards of the track, and I think on the north side," he wrote in April 1909 to his nephew Traven Aitken, who lived in Calgary. "I would

Bow River flooding, Calgary, 1908. TRANSALTA COLLECTION

think it worth your while to cover that district, with a view to ascertaining if there are any other likely power locations."

Max Aitken was one of the most colourful, determined, and controversial figures in the history of Canadian finance and business. From humble beginnings in New Brunswick, he achieved his first successes in Halifax before striking it rich in Montreal, the financial capital of Edwardian Canada. He was both a visionary risk-taker and a shrewd judge of business opportunities. He was also a ruthless consolidator and created two near monopolies, Canada Cement and the Steel Company of Canada. These struggles for market dominance made him a very wealthy man, but left behind a string of bitter critics who had lost out to his forceful business tactics.

Aitken's entrepreneurial endeavours in the Alberta electricity business stirred up far less controversy. Given the nature of the business, and the fragmented prairie market of numerous small communities, a power monopoly was neither possible nor worth the effort. Aitken had already invested in electrical generation plants in Quebec and put together a series of successful power, lighting, and street railway utilities in the Caribbean.

From the start, Aitken had a simple but effective business plan: control all the best Bow River hydro sites and, with this cost-effective supply source, dominate Calgary's power supply. This would provide leverage on the future economy of the region as well as cheap power for the Canada Cement plants he owned at Exshaw and Calgary. In turn, he could access his Montreal assets for the financing and the technology to build the power generation facilities. It was a neatly integrated model involving people, capital, technology, and markets.

His financing model followed a pattern that he had employed many times. He knew how to balance debt and equity to retain control over the operation. Most of the capital was raised through bonds floated by Royal Securities and its links to the money markets in Montreal, New York, and London. Many of the bondholders were repeat clients of Aitken's from earlier projects. While the bonds were widely distributed, a small circle of friends, including key partners in the Royal Bank, the Bank of Montreal, and National Trust, closely controlled the equity voting shares.

In 1909, Aitken started planning his business moves. He became president of Calgary Power and his good friend R. B. Bennett became manager of operations. Bennett began investigating the hydro sites and ascertaining who owned the rights to them. He also explored the likelihood of a major power sale to the City of Calgary. The parties were quietly consulted, but things moved slowly. Aitken was preoccupied with his much larger cement and steel projects. But he never lost sight of his hydro plans, which he saw as part of a larger economic empire in the booming western province.

Aitken also had his nephew Traven explore other business opportunities. Traven attempted to purchase the city-owned street railway company for another Aitken company, Montreal Engineering. The failed attempt created some tension. The younger Aitken reported to his uncle that council was "too socialistic" and "grafters of the meanest kind." He abandoned the street railway acquisition, because he needed to keep the goodwill of council for any future power contract.

Though Aitken had access to several attractive dam sites along the Bow River close to Calgary, he turned his attention to the one with the most obvious potential. Horseshoe Falls, about 85 kilometres west of Calgary, was still within easy range of the Calgary market. W. M. Alexander and W. J. Budd owned the rights through their Calgary Power and Transmission Company. What made Horseshoe Falls even more attractive was that Alexander and Budd had negotiated a contract in 1909 to deliver power to the City of Calgary. They were rumoured to be in financial difficulty as a result of construction cost overruns and had transferred the rights to their engineers, C. B. Smith and W. G. Chace. Aitken, who had a knack for turning failing businesses into successful companies, offered to purchase the rights

William Maxwell Aitken

Max Aitken, 1909. CORBIS 42-19065988

Max Aitken was Calgary Power's first president—before the company even existed! He was the man behind the early development of the Horseshoe Falls hydro station that would supply the company's power in the early days.

Born at Maple, Ontario, in 1879, his family moved to New Brunswick when Aitken was a child, where he came to know and admire R. B. Bennett. Aitken moved to Calgary briefly in 1898, where he renewed his friendship with Bennett. For a time he worked in a bowling alley, and then for the man who would be both president of Calgary Power and prime minister of Canada.

But Aitken really got to know the business world by working as a secretary for John F. Stairs in Halifax. Aitken eventually became a partner with the older man, and they formed Royal Securities Corporation Limited. Aitken took over leadership of Royal Securities when Stairs passed away, and went on to help reorganize the Canada Cement Company in 1909. That same year, he also became president of the predecessor to Calgary Power Company Limited, which was registered in Alberta as a "foreign" company on 17 March 1911—foreign because it was headquartered in Montreal.

His presidency only lasted until 27 April 1910, when he turned over the leadership to Herbert S. Holt, who in turn resigned so that Calgary resident R. B. Bennett could become president of Calgary Power on 3 August 1911. Aitken's varied interests often took him away from Canada. He was active in utilities far and wide, owning Canadian and Latin American power, light, and tramway companies. He also purchased several newspapers in the United Kingdom, including the *Evening Standard*, the *Sunday Express*, the *Daily Express*, and the Glasgow *Evening Citizen*.

In 1911, at age thirty-two, he was knighted, and when he entered the House of Lords in 1918, he chose the title First Baron Beaverbrook, First Baronet of Beaverbrook, New Brunswick, and Cherkley, Surrey, England.

The millionaire businessman was also a philanthropist. He gave generously to the people of New Brunswick—an art gallery, an ice rink, a town hall, a theatre and parks, and several scholarships. He served as the minister of Aircraft Production during WWII for the United Kingdom and was a close political ally of Winston Churchill. Lord Beaverbrook died in 1964 in Surrey, England.

> **An Early Employee**
>
> "April 1910, I started with Calgary Power when I first came out here to Morley," recalled Bob Armistead in the early 1980s. "We got off the CPR and me and another fellow walked down to Horseshoe. I went down and asked the foreman how the chances were for a job. He asked if I had a profession, and I said I was a coal miner and I just came up from Pennsylvania. The foreman said I could start in the morning. When I started at Morley I had just enough money to buy some overalls, a blanket, and a plug of tobacco. Once I started to work I was getting twenty-five cents an hour for ten hours work, and we were paying twenty dollars a month for board."
>
> Bob helped build the first transmission line; it took a whole year, and they camped out all winter. And he was there on Sunday, 21 May 1911, when Calgary Power turned on the electricity for the first time. "Do I ever remember that day—just as though it was yesterday. I was twenty-nine years old then."

and allow Smith and Chace to continue as the project engineers—a decision he would later regret.

After further planning and design work in 1909, construction resumed at Horseshoe Falls in early 1910. Low winter flow rates exposed much of the riverbed for foundation work. The workers had to divert the river with crude wooden forms and spillways, and the foundations had to be anchored into the bedrock to ensure the dam could handle the force of spring runoff. The Horseshoe Falls site was a narrow valley with falls and rapids descending in a curve. It provided tough challenges for the construction team.

The early weeks of construction proceeded smoothly under the supervision of Smith and his colleagues. Then, in May 1910, disaster struck. Spring runoff arrived suddenly, with greater volume and force than anticipated. The torrents of water washed away wooden forms, cofferdams, spillways, and the incomplete foundations. The engineering plans had seriously underestimated the volume of the flow through the site. Work was suspended and the damage assessed. Bennett and Aitken were furious. They fired Smith and company and hired new engineers from Montreal Engineering. This was the beginning of the close partnership between Calgary Power and Montreal Engineering that was to endure until the 1990s. Construction resumed weeks later, and most of the work had to be redone. As a result, the project was behind schedule and over budget. And in early 1911, a smallpox epidemic broke out in the construction crew, delaying completion of the project.

The faulty estimates of river flows proved to be a serious miscalculation during the construction at Horseshoe Falls and again, two decades later, at the Ghost Dam project. By then there was another side to the problem. Winter flow rates had also been miscalculated. As a result, Calgary Power overestimated the electricity it could provide when it negotiated the city contract and had to supplement its inadequate winter hydro-generating capacity with coal-fired generation at a plant in Calgary.

Construction was only one side of the business. R. B. Bennett had been hard at work renegotiating the 1909 Alexander and Budd contract with the City of Calgary. It called for two thousand horsepower with a provision for extension by increments of five hundred horsepower to a maximum of three thousand horsepower. The power was to be delivered beginning 1 April 1911 and continue through 1912. The basic price was $30 per horsepower per year, and the rate dropped to $28 for all power beyond two thousand horsepower. Aitken wanted a longer-term contract, but the city would not budge. Aitken hoped that once he had a contract, the city would gradually come to depend on Calgary Power.

The benefits of hydro power were soon evident to City council. Hydro power cost about half the price of the power from the city's Atlantic Avenue coal plant. This was a significant cost-of-living benefit

Two views of Horseshoe Falls hydro plant, c. 1911. TRANSALTA COLLECTION

to Calgarians as well as an incentive for new industrial development in the Bow Valley rather than near Edmonton. These lower prices helped quiet, at least temporarily, the calls for public ownership of the electrical generation system in southern Alberta.

To finance the work, Aitken raised about $3 million in debt and equity through the Montreal, New York, and London money markets, the major component being first mortgage bonds issued by Royal Securities and floated in London in 1910. By 1911, when electricity from Horseshoe Falls began serving Calgary, Aitken had left Montreal for London, and turned the presidency over to Herbert Holt, a close financial associate in Montreal. However, Aitken's letters to Bennett show his continuing direct involvement through 1913, and lesser role until he sold Royal Securities in 1917.

The Aitken-Bennett partnership had been the foundation for the successful launch of Calgary Power. The two men were very different in personality, values, and lifestyles. In his memoirs, Aitken summarized these differences: "He was noted for rectitude and godliness, I had a reputation for mischievous conduct." There were times when Bennett felt used by Aitken. In 1914, Bennett wrote bitterly: "I really do not propose to continue my business relations with your corporation … Definite promises involve definite results." However, the two men always patched up their differences once Aitken realized that he had pushed too hard. Without this two-man team, the company that was to become Calgary Power and the century-old TransAlta would not have happened. Aitken provided the financing and strategy, and Bennett delivered on local execution and political relations.

Herbert Holt had a very different style from Aitken. While the founding president was brash, outgoing, and aggressive, Holt was quiet, secretive, and avoided publicity. Nevertheless, he was one of the most powerful financial figures in Canada, though Calgarians today probably connect him with the luxury department store, Holt Renfrew. His involvement in Calgary Power was limited to strategic

The Beginning: 1909 or 1911?

The date that Calgary Power began operating as a corporate entity is confusing: some claim it was 1909 and others date the company to 1911. What happened?

In 1909, Max Aitken began work on both a street railway franchise and a hydroelectric plant on the Bow River. Calgary Power was formed as a company in Montreal in 1909, but without assets. In June 1910, the Montreal businessman purchased the Calgary Power and Transmission Company, which had started and then suspended construction of a hydroelectric plant at Horseshoe Falls. On 4 January 1911, the first active board of directors took over, and on 17 March 1911, the Calgary Power Company Limited was registered in Alberta as a "foreign" (Quebec-based) company. For many years, celebrations of the company's birthday took place on 21 May. On that date in 1911, Mayor J. W. Mitchell "put in the plug" to connect the city electrical grid to the hydro-power facility at Horseshoe Falls on the Bow River.

Calgary Power, later TransAlta, has been in the business of serving the people of Alberta—and beyond—ever since.

The Alberta Certificate of Registration of Calgary Power, signed 17 March 1911. TRANSALTA COLLECTION

overview; he left operational matters to Bennett, who took over as president in 1911.

The first power from Horseshoe Falls was delivered to Calgary through the new transmission line on 21 May 1911. There was no big celebration because time after time dates for the connection had been announced, then delayed. But the importance of the event did not go unnoticed in the *Calgary Herald*'s editorial page.

CALGARY'S NEW POWER

The turning on of Calgary's hydroelectric power yesterday was an event of immense importance to the city. It marked a new era in its industrial development. Today, Calgary is released from coal-made power. She has stepped into the ranks of those fortunate cities to which the streams and waterfalls pay tribute. The result can hardly yet be foretold, but one thing is assured— that Calgary's new standing in this respect will lend an immense impetus to the establishment of manufacturing concerns within her borders ... We may hope before long to have facts for submission to manufacturers, showing that Calgary, in the cost of power and in the cost of living, is unsurpassed as a field for industrial effort.

Calgary Power was officially in business to supply customers in Calgary. The new supply eventually allowed the city to close its Atlantic Avenue power plant and move the equipment from it to a more modern facility in Victoria Park. For now, the city wished to keep a certain amount of generation in its own hands—to give it some leverage over Calgary Power and as a backup in case of a power failure.

With electricity flowing through the wires to Calgary, Bennett, as Calgary Power's president, had to assess the chances of a rival emerging to challenge the young company. Any serious competition could threaten Calgary Power, which only had a two-year contract

continued on page 35

POWERING GENERATIONS

Herbert Holt

Herbert Holt, c. 1910. TRANSALTA COLLECTION

Born in Ballycrystal, Ireland, in 1856, Herbert Samuel Holt came to Canada in 1873 and as an engineer worked on some of the biggest railway construction projects of the time.

He worked on the difficult section of the CPR line west of Calgary through the mountains and helped complete the transcontinental line in 1885. Holt also helped build the Calgary and Edmonton Electric Railways, later went on build more rail lines elsewhere in Canada and in the United States, and moved on to South America where he built the Trans-Andean railroad. By 1915, the year he was knighted by King George V, Holt had been involved with the Montreal Gas Company, the Montreal Park & Island Railway, the Montreal Light, Heat & Power Co., the Imperial Typewriter Machine Company, the Dominion Textile Company, the Canada General Electric Company, the National Trust Company, the Montreal Trust Company, the Canadian Car and Foundry Company, the Royal Electric Company, the London Street Railway, and the Monterey Railway and Light Co., among many other ventures. Under his leadership as its president from 1908 to 1934, the Royal Bank of Canada became the first billion-dollar bank in the Dominion. He served as the bank's chairman from 1934 until his death in 1941.

Holt became a director and then president of Calgary Power in 1910. On 3 August 1911, after power began to flow from the Horseshoe Falls hydro plant to Calgary, he wrote a decisive letter to the directors of the recently formed Calgary Power Company Ltd.

"Now that the construction work is practically completed, I consider that it is vitally important for the welfare and interest of the company that the chief executive officer should reside in the West. I have therefore asked Mr. R. B. Bennett if he would accept the office of President which I am pleased to say he has consented to do." Holt was a smart enough leader to know when to resign. He turned over the reins of the exciting new enterprise to an influential Calgary lawyer and man-about-town. Though the headquarters of the company did not move to Calgary until 1947, Holt's prescient resignation paved the way for Calgary Power to become an institution firmly rooted in the west.

A great believer in personal initiative, the capitalist Holt responded to criticism in the 1930s by saying: "If I am rich and powerful, while you are suffering the stranglehold of poverty and the humiliation of social assistance; if I was able, at the peak of the Depression, to make 150 percent profits each year, it is foolishness on your part, and as for me, it is the fruit of a wise administration."

Upon formation of the Canadian Business Hall of Fame in 1979, Holt was recognized for his outstanding achievements and contributions to Canadian business.

Victoria Park plant, Calgary, 1911. TRANSALTA COLLECTION

with the City of Calgary. Calls for municipal ownership of the electrical generation system continued, but the city had moved too late to secure the Radnor hydro site on the Bow River near Cochrane. In December 1910, the city had applied for and received a conditional federal licence for the Elbow Falls site on the Elbow River west of Bragg Creek. Bennett never took this challenge seriously, because winter flow rates on the Elbow were too low; however, he was concerned about a private sector proposal to build a series of four dams on the Bow River downstream (east) of Calgary. The timing could not have been worse for this proposal, as the financial markets slid into a recession in 1913 and World War I broke out in 1914. As a result, Calgary Power became the dominant—and only—supplier of electricity to the Calgary market.

By the time the Horseshoe Falls plant was completed in 1911, Calgary Power's engineers realized that the winter flow rates would never generate enough power to fulfill the Calgary contract. There were two solutions: more storage above the Horseshoe Falls dam, or more generation. Storage would require approval by Parks Canada, which was unlikely, so the company decided to build more generation by seeking the rights to develop the Kananaskis Falls site, just above the existing Horseshoe Falls plant on the Bow River.

Bennett's efforts to secure the rights to develop the Kananaskis Falls site encountered some unexpected opposition. Dr. Andrew Macphail, the first professor of the history of medicine at McGill and a practitioner at the Montreal Hospital for the Insane, was the unlikely applicant and rival. He was well known for his articles attacking the business practices of men like Max Aitken. Along with some business associates, he made a formal application, complete with technical drawings, for a hydro facility at the site. But Macphail was a Liberal, and, when Robert Borden and the Conservatives swept into power in 1911, Bennett was able to use his connections in Ottawa to Calgary Power's advantage. "I have succeeded in getting the Department of the Interior to give us Kananaskis Falls," wrote Bennett to Aitken in early 1912, "and will have the lease issued in due course." It also made sense to have one company operate both sites near the confluence of the Kananaskis and Bow Rivers.

Unfortunately for Bennett, the complications were not confined to Ottawa. The site included land owned by the Stoney Band, and the project affected the band's aboriginal fishing rights and called into question financial compensation (from the federal government) for land claims. Negotiations between Calgary Power, the Stoneys, and Ottawa's Indian Department officials dragged on. After several years, with the plant nearing completion, the Stoneys decided on a more aggressive strategy. The crisis came to a head in May 1914 when rumours arose that the band was preparing to destroy the dam and the plant. Company officials demanded police protection for property and workers, given the "grave danger of trouble." Ottawa intervened to prevent violence. The federal government invited the Stoney chiefs to Canada's capital, where they met with company and government officials and hammered out an agreement. The band

Construction of Kananaskis Falls dam, 1913. TRANSALTA COLLECTION

An Italian Immigrant and Calgary Power

"Fourth of March, 1911, I come here from Italy," John Strappazzon recalled in 1980. "I got a ticket to go to Exshaw. I am twenty when I come over. I get off at half past nine at night. Cold—no dress up like Italy. No overcoat, mitts, nothing. I cold like the dickens. 'You have to wait until spring,' they told me. So I ask your fellow there, 'How far is down there to Horseshoe?' 'Oh, about ten miles.' I get valise, wrapping up, and I find a stick and I go down the track, walking with my valise, alone."

And that's how an Italian immigrant got hired on with Calgary Power, where he remained for his entire working career.

John Strappazzon (right) at Bearspaw Dam construction site with Johnny Bearspaw, Chief David Bearspaw, and Walking Buffalo, 1954.
TRANSALTA COLLECTION

---

members received $25 each for settlement of their claims. The band got $9,000 for the purchase of ninety-four acres of land, and $7,500 per annum for water rentals to 1918. The agreement defused the impasse, and sensitivity concerning aboriginal rights and aspirations became a critical component in future project development.

One further complication arose during the construction of the Kananaskis Falls project. The dam flooded upstream to cover about forty hectares of Rocky Mountain National Park—today's Banff National Park. The issue was resolved quietly, but relationships with the authorities in the national parks became an increasingly important factor in hydro development further west on the Bow River.

Compared with the difficulties involved in building the Horseshoe Falls dam and power plant, the Kananaskis project went relatively smoothly. Montreal Engineering supervised the design and construction of this project. Careful planning and management avoided the spring runoff problems. The work force was larger this time, about four hundred to five hundred men, making construction advance more quickly. The workers took great pride in the strength and permanence of their structure, which had been built using manual labour and horses.

Lack of water storage was a continual problem. The peak flow was in June, while the peak market was from December to March. The peak could be as much as twenty times the minimum volume. Full revenue from the plants could never be achieved if the company relied on natural water flow rates. One solution was to capture and store the water at a point above the two power plants and release it when the electricity was needed. Calgary Power and the City of Calgary also agreed to use the Victoria Park thermal generating station to complement the hydroelectric plant outputs and make up the difference.

Bennett knew the solution to the storage problem lay in expanding Lake Minnewanka so that it could hold more water for winter power generation. However, the lake was in the national park, near the town of Banff, which created a series of conflicts with the National Parks Act and park policies. Minnewanka was a favourite summer spot for local residents. Some of them had built cottages on the lakeshore. There had been a small hotel and tour boats at the lake since the 1890s. In 1908, Parks had built a small 1.2-metre dam to improve navigation on the lake. In 1912, after extensive negotiations, Calgary Power received permission to raise the height of the dam to 4.9 metres. Unfortunately, the water flow was still lower than the company had projected. As a result, in 1914, the company went back to Parks seeking a further expansion of the dam. This time, Calgary Power met stiff opposition from both Parks and Fisheries officials, who had been strongly criticized for the previous modification. The cottagers wanted no further changes in lake levels. Although the outbreak of World War I in 1914 postponed the Minnewanka project, Calgary Power still needed more water storage in order to serve its customers.

While the two dams on the Bow were under construction, global economic conditions were changing dramatically, and the Calgary economic boom went bust. Capital markets dried up, consumer spending dropped, and the demand for electricity levelled off. The market for grain and cattle plummeted, and the huge stockyards in Calgary felt the effect. The city negotiated lower electricity prices when it signed a second contract with Calgary Power. The price fell from $30 per horsepower per year to $26.

By the end of its first five years, Calgary Power was managing a complex series of stakeholder relations including Calgary City council, the Stoney band, national parks officials, an often critical media, and local business owners who lobbied via the chamber of commerce for cheaper power. Nevertheless, Bennett believed that as long as Calgary Power delivered reliable power at a fair price, the company's future would remain secure.

Gifford Horspool and son, c. 1928. Horspool possessed exceptional mechanical skills and worked for Calgary Power from 1920 to 1964. LUCILLE ROXBURGH

A. E. Cross, Rancher and Calgary Power Leader

A. E. Cross, 1915. GLENBOW ARCHIVES S-222-17

At first glance he did not seem to fit with the others, but the rugged rancher was an important part of the mix in a board of directors that assured Calgary Power's success. Alfred Ernest Cross, born in Montreal in 1861, and educated at both the Montreal Business College and a veterinary school, arrived in Alberta in 1884. Just twenty-four years old, he started the A7 Ranch west of Nanton in 1885. It was the largest ranch in the territory.

Cross founded the Calgary Brewing and Malting Company in 1892, and was its president until his death in 1932. He served as a member of the North-West Assembly from 1899 to 1902, before Alberta was formed in 1905. He was a founding member of the Western Stock Growers' Association and the Calgary Board of Trade. He helped start the Ranchmen's Club and served as its president from 1906 to 1908 and again from 1911 to 1912. Cross was one of the "Big Four" who founded the Calgary Exhibition and Stampede in 1912. The rancher also helped start Calgary Petroleum Products Company in 1912—the outfit that discovered oil at Turner Valley in 1914—and he was a director of Canadian Western Natural Gas.

According to the late Henry C. Klassen, professor of history at the University of Calgary, Cross "compares with [Patrick] Burns and [Alexander] Galt as an agent of modernization, and his name symbolizes the forces of growth in both the regional and national economies; a spirit of enterprise, the normal pursuit of profits, family capitalism, access to central Canada's and Britain's capital markets, and economic progress through reinvestment of earnings."

A. E. Cross was a Calgary Power Company director until his death. "Mr. Cross had been a Director of the Company since its inception in 1909," noted a tribute to him in the company's 1932 annual report, "and rendered valuable services in connection with the conduct of its affairs." His son, John Braehead Cross, served on the board of Calgary Power from 1956 to 1979, continuing the tradition of ranchers helping direct the operation of the utility company.

The War Years

In the lazy summer days of August 1914, Albertans were focused on the joys of family and holidays when shocking news arrived from Europe. Gavrilo Princip had assassinated Archduke Franz Ferdinand, heir to the throne of the Austro-Hungarian Empire, and his wife, Sophie. The murder precipitated the outbreak of World War I. In spite of the surprise, Canada mobilized quickly in support of Britain and the Empire. As ardent patriots, Max Aitken and R. B. Bennett were emphatic that Calgary Power support the war effort, and employees who enlisted were guaranteed a job when they returned.

Canada was in a recession when the war began. The conflict raised expectations that government purchases would jumpstart the economy. But orders for food, horses, and uniforms did not end the hard times. The major orders—for arms and equipment—went to companies in Ontario and Quebec, frustrating western politicians. The continued recession also limited the demand for electricity.

Like other businesses, Calgary Power faced major challenges during the war. Replacements had to be found for employees who had enlisted. Security at hydro plants and isolated transmission lines became a serious concern. Social tensions flared up in the work force as immigrants from Central Europe found themselves mistrusted by colleagues and treated as if they were agents of the Kaiser.

Those Calgarians who lost loved ones in the fighting suffered deep psychological scars as did those who waited month after month, year after year, for that dreaded telegram from Ottawa. Through all this, Calgary Power employees responded with enormous effort and dedication. There was a willingness to adapt, and a spirit of collaboration and commitment kept the essential power flowing. To this day, every November, on Remembrance Day, the company exhibits the portraits of its employees who died in battle.

But not all of the battles took place in the trenches in France. Corporate boardrooms faced their own skirmishes as a result of the

Victory Bond drive during WWI, Calgary. GLENBOW ARCHIVES PA-3476-16

war. Large international trusts or holding companies with a huge portfolio of companies came into existence during the war. In 1915, General Electric, through its utility portfolio, proposed a takeover of both Montreal-based Royal Securities and of Calgary Power. Izaac Walton Killam, investment guru and president of Royal Securities, investigated the offer for Aitken, who was in London at the time, though still the principal shareholder in both companies. The takeover failed. The shareholders believed they had a much brighter future ahead as an independent Canadian company than as an arm of General Electric, in spite of the temptation of quick profits.

During the company's first decade, company president R. B. Bennett had used his political skills to promote the private enterprise option offered by Calgary Power. In 1915, Alberta premier Arthur Sifton created the Public Utilities Board (PUB), but the regulatory powers of this provincial agency did not extend to the contracts between Calgary Power and its customers—Cochrane, Calgary, and

Victor Drury

Victor Drury, 1920s
TRANSALTA COLLECTION

The early 1920s were years of recession, and the late 1920s saw economic growth. So Calgary Power was changing rapidly when Victor Montague Drury took over as president of the company in 1921. A banker, he started with the Bank of Montreal and in 1909 moved to Montreal Trust and later to Royal Securities, where he became president in 1919.

As president of Calgary Power from 1921 to 1924, he was involved in renegotiating the company's contract with the City of Calgary in 1923, as well as the company's quest for more hydro capacity along the Bow River. The company also began construction of the third transmission line from Seebe, in Exshaw, to Calgary during his period of leadership, which allowed for the expansion of service to other communities in southern Alberta later in the decade.

In 1924, he left to start his own businesses, which included the Canadian Car and Foundry Company and the Provincial Transport Company. He was also involved with the Eddy Match Company and Sherwin-Williams Paints. An active member of the Calgary community, he was an honorary member of the Canadian General Council of Boy Scouts and a benefactor to orphanages.

---

the Canada Cement plant—or to the rates charged by the utility. Indeed, electrical utilities had to submit their contracts to the PUB for approval only after they had negotiated them with their clients. Nevertheless, the creation of the PUB by the province was the first of a series of regulatory agencies that would eventually become part of the Calgary Power story.

A New Partnership

The first decade of Calgary Power bore the stamp of Max Aitken and R. B. Bennett. And the years following the war would become known as the "Killam years," as Izaak Walton Killam took over the direction of the company, following the brief presidency of Victor Drury. Killam had become president of Royal Securities in 1915 and joined the board of Calgary Power in 1917. He bought out Aitken in 1919 and became company president in 1924. He carefully picked his able successor, Geoffrey Gaherty—a brilliant young engineer and hydro enthusiast. Killam and Gaherty developed a very effective working relationship somewhat similar to the one shared by Aitken and Bennett.

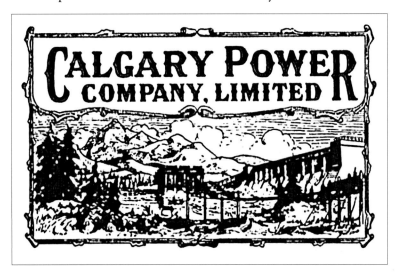

Calgary Power logo, 1920s–30s. TRANSALTA COLLECTION

POWERING GENERATIONS

Survey crew at Spray Lakes. L to R: Unidentified, A. E. Roebotham, Scotty Lyall, and Bill Wolley-Dod, c. 1920. TRANSALTA COLLECTION

The city contract came up for renewal in 1923. As a result, the price dropped to $15 per horsepower per year—exactly half of the original price charged in 1911. Better equipment and higher efficiency allowed the company to make larger profits each year during the 1920s while lowering its charge to its customers. Equally important was the integration of the Victoria Park coal-fired generating plant, which Calgary Power gained use of when it bought out the shares of the old Calgary Water Power Company and its lease of the city-owned facility in 1928. Gaherty noted that this development resulted "in our having full control of the power supply in the city of Calgary." The coal-fired facility was turned on during times of high electrical demand to supplement the steady power from the hydro units on the Bow River that were operating at maximum capacity.

Fortunately for Killam, the economy rebounded in the years to follow and caught up with the company's generating capacity, spurring on the quest for new generation. Calgary Power employees explored the wilderness south of Canmore, looking for hydro sites. The company considered the Spray Lakes area, but development of this region came much later due to complex economic and political reasons.

Lake Minnewanka remained an option. In 1921, Calgary Power applied once more to the federal government for permission to increase the dam and storage capacity of this reservoir. When local residents and cottagers again opposed the plan, the administration in Rocky Mountains Park rejected the application. In 1922, Parks closed the coal-fired power plant at Bankhead, and then installed a small hydro plant at the existing dam on Lake Minnewanka to supply electricity for Banff.

As a result, Calgary Power turned its attention to its long-delayed Ghost River project site. Just 45 kilometres west of Calgary at the junction of the Ghost and Bow Rivers, "the Ghost" filled a wide valley that offered significant storage potential with a head of 33 metres. But it required a 1,500 metre-wide dam with 275 metres of poured concrete flanked by earth-filled walls.

Calgary Power used new steam technology to build the Ghost Dam—no more pick, shovel, and draft-horse teams. The two-year project employed 150 men and two giant Bucyrus steam shovels to excavate 4.5 million cubic yards of fill using seven trains of dump cars. The fill had to be transported 3.2 kilometres to the dam site, and the project required the construction of high-level trestles from which to dump the sand and gravel. Even so, spring runoff carried away construction forms and equipment. When the reservoir filled, it created a lake 12 kilometres long and more than 1 kilometre wide. Calgary Power installed the first of two 1,800 horsepower units there in October 1929.

During the 1920s, Calgary Power expanded its market network dramatically. In 1926, it moved south into the High River area, where it lowered the cost per kilowatt-hour from 18 cents to 9.5 cents. Lethbridge turned on to Calgary Power in 1927, and in the remaining years of the decade, dozens of other communities signed contracts with the utility. In 1928, the company moved north, migrating toward the communities around Edmonton, and hooked up Wetaskiwin. By 1930, Calgary Power boasted a transmission system that stretched from the U.S. border to 80 kilometres north of Edmonton, reaching 136 communities and 250,000 people.

Calgary Power's first two decades were successful for a variety of reasons. The booming west provided an unparalleled opportunity for a new utility company with the right resources, talent, and connections. Calgary's boom years from 1900 to 1913 had created a market demand for new power generation that could not be served by the small coal-fired units. And the deep pockets of Royal Securities and the technical hydro expertise of Montreal Engineering gave the upstart company a definite edge over the competition. Calgary Power

(top left) *Ice breaking up, with suspension bridge, during construction of the Ghost plant, 17 November 1928.* (top right) *Ghost construction, 1928.* (bottom) *Completed Ghost Dam, 1929.* TRANSALTA COLLECTION

seized these opportunities and grew from a good idea into the principal supplier of power to the City of Calgary.

The company's extensive generation and transmission system, financed from within the Montreal-based corporate family, allowed Calgary Power to provide low-cost electricity to more of the province. And as prices came down, the demand for public ownership eroded. The company considered numerous hydro-power sites and developed the prime candidates, along with storage reserves, in time to meet the electrical demands of a province desperate for power in the 1920s, just as the Killam-Gaherty partnership stepped in to lead the company.

In his English study, Aitken must have read the annual reports from Canada and realized that the foundations of the company had truly been built to last. With the Great Depression and another world war on the way, Calgary Power would need this solid underpinning to survive the difficult years ahead.

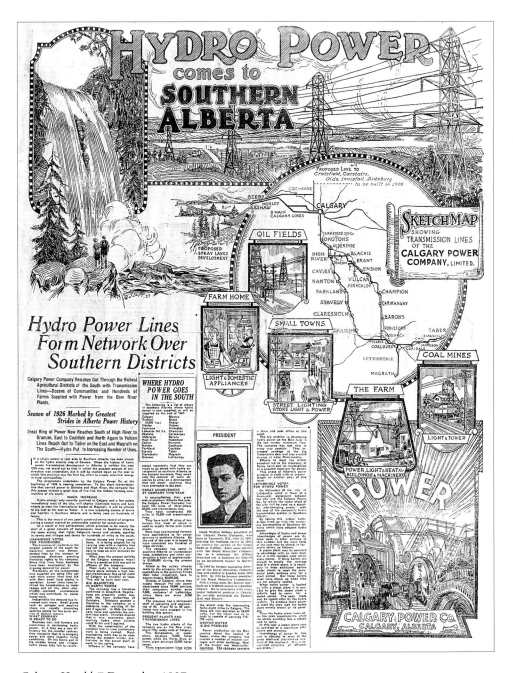

Calgary Herald, 7 December 1927

"To this day, I take a little pride when I drive through the country at night, and see all those yard lights shining from all those farms, all over the place, and I remember buying those poles and those mercury vapour lamps—I always feel kind of a part of that. That's my heritage." DEAN HAGGINS, CALGARY POWER EMPLOYEE, 1930s

CHAPTER FOUR

Depression, War, and Survival

The Dirty Thirties were a tough decade for Canada and the darkest days in Calgary Power's history. The prosperity of the 1920s had triggered Izaak Killam's expansion plans, but severe drought crippled the agricultural sector, and the stock market crash in November 1929 brought down the rest of the economy. The downward spiral continued for the next four years as markets attempted to find bottom. The heavy investments in the late 1920s to build the Ghost Dam—and double the company's generation capacity—had created a risky debt load just as markets fell and prices declined.

With a sense of helplessness, many Albertans turned to government for support. In 1930, R. B. Bennett's Conservatives defeated Mackenzie King's Liberals and took power in Ottawa. In Edmonton, the provincial treasury was ill equipped to deal with a severe international crisis, and the government, led by the United Farmers of Alberta party, was overwhelmed by events. The economic collapse cut revenues just when the demand for public services peaked. Drought-stricken farm families fled from the land in search of city jobs—but no jobs existed. Long bread lines challenged the social agencies and churches that tried to feed the hungry.

Calgary Power did not supply the rural areas of the province, where the situation was the worst. The demand for electricity did not drop very much in the urban areas, and it fell to the City of Calgary's electricity department to collect on the accounts of customers who

Ray Boissoneault, John Strapazzon, A. W. Moore, and Bob Brownie at Kananaskis power plant, 1941. TRANSALTA COLLECTION

The unemployed demonstrate in Edmonton, December 1929. GLENBOW ARCHIVES NC-6-12400A

could not pay their bills. Calgary Power's ten-year contract with the city provided stable income, and the company made up for declining sales by seeking new customers.

The Killam-Gaherty Team

Izaak Walton Killam and Geoffrey Gaherty would successfully guide Calgary Power's operations throughout the hard times of the 1930s and carefully grow the company in the early 1940s. They expanded it beyond Calgary into southern Alberta and prepared for a time when the utility company could become the dominant player in Alberta.

As head of Royal Securities, Izaak Killam was central to this financial institution, a maker of shrewd deals, and one of the most talented financiers in Canadian history. Unlike Aitken, his successor was quiet-spoken, introspective, brooding, and secretive. Killam disliked public appearances, speeches, or even having his picture taken. His idea of relaxation was salmon fishing in New Brunswick or sitting in a quiet corner in one of his Montreal or New York private clubs.

continued on page 52

Hap Hansen (right) hands out prizes at the Calgary Power company picnic in Bowness Park, Calgary, c. 1954. VIC JONES

Darrel "Hap" or "Happy" Hansen

Many communities with their own power plants were hesitant to sign electric franchise agreements with Calgary Power. But Hap Hansen became famous for the way he masterfully negotiated with these important clients during the 1930s and beyond, helping Calgary Power expand its customer base even during the hardest of times. He recalled his strategies for winning them over when interviewed in 1983:

"I found that members of the [town or village] council, who were just normal citizens in each community, mistrusted the company's efforts, and I decided on a philosophical way that I thought might work, which was that they might not trust the company until they trusted its representative.

"So I met with individual councillors, members of the staff in each town as frequently as I could, discussing our problems as well as their own at each visit until I found that as they gradually understood the problems of our particular company, of generation, transmission, distribution, and delivery of power in those pioneer days, I think as they began to understand it better, they began to trust us better.

"I'll never forget a council meeting in one of the towns east of Camrose. It was a renewal period, and I spent some time in preliminary discussion with them, but the secretary-treasurer began interrupting many times, saying Harry Thompson was a crook, Geoff Gaherty was a crook, and even Mr. Killam was a crook. So finally I said to the mayor, 'We're not getting anywhere.'

"I closed up my file and my briefcase, put on my hat, and walked out. The mayor followed me and said, 'Come on back, I'll explain to the secretary that this shouldn't go on.' In a few minutes, they'd given the necessary bylaw two readings for public utilities' approval, and we all went over to the hotel for a drink by the secretary. Next morning, I went back to apologize to the secretary, and we became close friends."

Izaak Killam, c. 1925. TRANSALTA COLLECTION

Izaak Killam

The man who was rejected from service in the Great War due to a "tricky heart" went on to become one of the most compassionate business leaders Canada has ever known. Always tight with money—his own as well as that of his companies—Killam's name is also remembered for his philanthropy.

Izaak Walton Killam was born in 1885 in Yarmouth, Nova Scotia. As a twelve year old he sold newspapers and by age fourteen had cornered the market on street sales in his community. His father died when Killam was sixteen, so he took over as head of the household. At age eighteen, he began his career as a clerk in a bank and the next year became a securities salesman for Royal Securities. He moved to the Montreal head office for the company in 1906, and by 1915 the thirty year old was president of Royal Securities.

Max Aitken described Killam as "truthful, upright, indispensable." The crash of a pulp and paper company in which he had invested in the 1920s made him more conservative, but during that decade he also invested in real estate, grain elevators in Western Canada, the Cuban Electric Co., tin in Bolivia, oil in Venezuela, and power in British Guiana, Puerto Rico, Mexico, and parts of South America. According to his biography, he "spotted potential, got in cheap and held on while he developed or expanded the business ... He was a builder." In Canada, Killam and Montreal Engineering—another one of his companies—developed power in Ontario, Quebec, Nova Scotia, New Brunswick, PEI, and Newfoundland.

Killam greatly appreciated the potential of southern Alberta and joined the board of Calgary Power in 1917. He became the majority shareholder in 1919 and chairman of the board until 1955. During his presidency of the company, from 1924 to 1928, Calgary Power expanded its system throughout Alberta.

The great visionary was principled. A young Marshall Williams—later to become president of Calgary Power—worked with Killam and said, "He always had the long-term view." Killam invested in electricity because it was a basic ingredient for many other industries. He believed in public transportation and built transit systems. And he built pulp and paper mills, in part to supply the newspapers that were included in his investment portfolio.

"We make money," Killam told Williams, "but our purpose is to serve the customers. Customers are number one. If you serve them well, the shareholders will do well." Harry Schaefer, one of three generations of Schaefer men who have worked for the company, learned Killam's mantra, too. "The motivating factor was, if you looked after the customer, the profit would look after itself," Shaefer said. It even applied in a most unusual case history. "There was an agreement signed, negotiated in Prince Edward Island, and he looked at it and said, 'It's too good. Too good for us; it won't stand the test of time.'" So Killam forced his people to go back and renegotiate the contract. "He was prepared to do the right business deal, not the one that got too much on your side of the table." Planning was important to the leader. "Just stop every day for ten minutes and think in the future tense," Williams recalled. "What if? What if? What if? His whole thinking was building for the future."

Killam's health suffered due to his work habits. In 1954, he stepped down from most of his business duties and sold Montreal Engineering to his employees. He had been an avid fisherman since the 1920s and bought land along the Restigouche River. As his health failed in the 1950s, he spent most of his time fishing. When his physician forbad him to go out, he took the doctor along in the canoe. A heart attack in 1954 only made him more determined. "Despite his heart attack earlier that year, he never fished more or harder than he did in 1954," writes Douglas How in Killam's biography. "He killed many salmon, big salmon, one of 42 pounds, one of 43, one of 45, friends say, and in keeping with the time-honoured ritual the silhouettes of all three were traced on boards which were hung on a wall for all to see. He was so proud of the biggest that he kept it intact until it began to stink." On 5 August 1955, after a morning out fishing, he died while taking a nap.

Calgary Power's 1955 Annual Report summarized Killam's accomplishments: "His foresight and tenacity of purpose put Calgary Power on its feet and brought it through the Depression and concurrent drought. His judicious planning made it possible for the company to withstand the impact of inflation and to finance the present rapid expansion without taxing its resources."

The Killam Scholarships, created by his widow, Dorothy, "help in the building of Canada's future by encouraging advanced study," and continue his legacy.

Geoffrey Gaherty, 1964. TRANSALTA COLLECTION

Geoffrey Gaherty

He was reluctant to be talked into staking the company's future on coal, but once convinced, the engineer and economist built an underpinning for the company during thirty-two years as president that still stands it in good stead to this day.

Geoff Gaherty really was a hydro man, from the day he hired on with Montreal Engineering in 1920. He quickly made his mark with Izaak Killam, who desperately needed able young executives to run his companies. Gaherty became a director in 1924, and during his tenure as president of Calgary Power from 1928 to 1960, nine hydro stations were brought on-line along the Bow River and its tributaries. According to long-time company engineer Jack Sexton, Gaherty was the most influential president of Calgary Power. "He lived that company and he lived its expansion. When he arrived on the scene after the First World War in the very early 1920s, it was a poverty stricken little company. The Bow River really is a very poor hydro river ... It took a man of his capabilities to see ... its potential to be adapted to Calgary's needs ... In its natural state it would drop down to two or three hundred cubic feet per second in the wintertime, a little stream."

"It was the vision of Geoff Gaherty and his nation-wide reputation to do with hydroelectric development that put Calgary Power in its excellent position for expansion," recalled Hap Hansen.

Gaherty was larger than life. A chain smoker, he always had a cigarette in his hand. And he never lived in Calgary, even though Calgary Power's operations were all in Alberta during his lifetime. He was president of corporate siblings Ottawa Valley Power and Montreal Engineering until his retirement in 1960. Once or twice a year Gaherty arrived by train from Montreal, and, according to Harry Schaefer, he would "set up camp in the Palliser Hotel. At the end of each of the towers they had a conference room with a fireplace in it." Jack Sexton recalled that Gaherty had an unusual habit. "One of the first things he would do when he entered his suite at the Palliser Hotel was to go around and turn on all the lights. I think that was more symbolic than anything else."

In 1947, the company was traded for the first time as an investor-owned utility, the same year its name was changed to Calgary Power Ltd., and its headquarters moved to Calgary. In 1948, the company formed Farm Electric Service Ltd. as a subsidiary to develop a power grid for rural areas, replacing the company's Rural Department.

After World War II, the federal government began giving tax credits to companies for donations to public institutions, Sexton recalled. "It encouraged corporations like Calgary Power to make donations to educational and charitable institutions, and Calgary Power decided to make a nice donation to the University of Alberta," he said. "I remember after the decision was taken, Mr. Gaherty, from Montreal, phoning the [general] manager, Harry Sherman, who was another delightful character. Mr. Gaherty was quite pleased with himself about this donation to the University of Alberta, and he started questioning Harry. 'Now what do the staff think about that?' 'Well,' Harry said, 'They would think an awful lot more about it if you would use the money to start a pension scheme!' And it was subsequently done."

When it became evident that no more hydro facilities could be developed on the Bow River, Gaherty began looking for opportunities elsewhere. "Yes, it was interesting working with him and, of course, he loved hydro," recalled Sexton. "He resisted the introduction of thermal power for Calgary Power. Once he came around, he was very brilliant because they started that big plant at Wabamun on gas generation. But he was the one that said gas is too valuable a fuel to burn under central station boilers. These boilers—the next units at Wabamun—are to be coal fired. And of course that is why it was located near the Wabamun coal mine."

Accustomed to running a group of companies as he saw fit, Gaherty did not like the increasing regulation and government control that emerged in the 1950s. As Harry Schaefer recalled, "He was pretty annoyed with all of this regulatory stuff and he said, 'We've got to repel the boarders!'"

Gaherty won many awards during his lifetime. The Stoney Nation west of Calgary named him Chief Powerful Waters for the work he did along the Bow River. And as a life member of the Canadian Electrical Association, he won the Julian C. Smith Memorial Award from that organization in 1947 for achievement in the development of Canada.

Gaherty and his wife with Stoney band members. Gaherty was honoured by the Stoneys for his good relationship with the band, 1960. TRANSALTA COLLECTION

Kananaskis Dam, c. 1940. GLENBOW ARCHIVES PA-3538-1

His intense desire for privacy created an air of mystery and intrigue that only added to his business reputation.

During the boom of the late 1920s, Killam had adopted an aggressive business strategy and positioned the company as the leading electrical supplier in Alberta. Geoff Gaherty, as company president, had led the campaign to sign electric franchise agreements with numerous small villages and towns in southern Alberta, purchased a series of municipal steam-powered generating systems, built new transmission lines, and serviced the expanding system with electricity from Calgary Power's hydro facilities. The resulting low power prices and twenty-four hour supply—which most small utilities did not provide—neutralized any local competition.

With the collapse of capital markets, the role of Royal Securities became pivotal to Calgary Power. The financial company provided some capital directly and provided collateral for securing debt from other lenders. The government also strengthened the company's hand. Federal officials allowed banks to value corporate bonds for collateral on the basis of their face value, not their market value—an arrangement that benefitted Calgary Power.

Calgary Power took many years to pay off the Ghost construction expenses and find markets for the excess power. It executed cost-cutting measures while Killam recycled his dividends back into the company. Somehow, all the payrolls were met and the liabilities covered. Killam's connections procured several bridge loans from the Royal Bank to get through tough periods, and in 1934 he lowered debt costs by converting a $1.8-million bank loan into bonds. The worst was over by 1935 as the markets began to stabilize.

While Killam provided the financial vision for the enterprise during the Depression, it was Gaherty who crafted and executed the business plan—both operating from Calgary Power's head office in

George Pocaterra and R. Malraison at Flat Creek in Kananaskis Country, c. 1920s. GLENBOW ARCHIVES NA-695-31

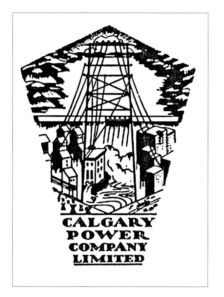

Calgary Power logo, 1930s.
TRANSALTA COLLECTION

Montreal. An excellent hydro engineer, Gaherty could envision the power potential of a river by just looking at a valley. He loved the technical challenges of harnessing turbulent rivers, and he was responsible for the second group of Calgary Power hydro projects—all still operating today. Gaherty's combination of dams, diversions, and flow patterns maximized power generation, and these projects delayed the need for thermal generation with natural gas and coal until the 1950s and 1960s.

Despite the constraints of the Depression, Killam and Gaherty continued to plan for growth. The change to Banff National Park's boundaries in 1930 provided Calgary Power with the opportunity to explore the region south of Canmore for hydro development. Reservoirs high in the mountains would benefit every power plant in the downstream watershed. Parties of rugged outdoor employees spent cold and lonely winters up in those valleys trying to estimate flow rates and establish potential reservoir capacity. Their winter exploits became part of the Calgary Power story. They regaled their colleagues with tales—some of them tall—of near starvation, falling through the ice, or cougar and grizzly bear attacks. The "heroic myths" of the early days added to the mystery and romance of hydro development.

Spring runoff from the extensive snow pack of the high mountain valleys suggested potential for new generation facilities and pointed to the valley bottoms that could be dammed for water storage. While major projects in the mountains were delayed until after the Depression, Calgary Power completed several other initiatives with funding from federal government make-work programs. One such project was the raising of Kananaskis Lake six metres, which increased its storage by 43 million cubic metres. During this difficult decade, Calgary Power engineers squeezed out efficiencies from existing plants.

continued on page 56

Seebe, 1909-2004

The name Seebe comes from the Cree word "si-pi," which means creek, river, or meeting of the waters. All these descriptors apply to the little hamlet at the confluence of the Bow and Kananaskis Rivers, a community that arose in the early 1900s to support the Calgary Power dams at that location.

In 1909, Calgary Power Company Ltd. acquired land from the Stoney reserve for Horseshoe Falls Dam and in 1913 for the Kananaskis Dam. The hamlet of Seebe sprang up during the construction of these dams and eventually included twenty-two houses, a seventeen-unit apartment complex, a one-room schoolhouse, a baseball diamond, and a curling rink. The *Henderson's Directory* for southern Alberta in 1914 lists Seebe as a community with 350 residents and mail delivery four times a day. In 1918, the community had enough children to support a one-room school, and a general store opened. Alberta School District 4152 was officially erected with Seebe as its name on 9 November 1922. "J. F. Boyce, B.A.," was the trustee. In a hamlet as remote as Seebe, the school not only served as a teaching institution for Grades 1 to 6, but also as the community centre. Weddings, parties, performances, and other events took place at the school and in the community hall next door. Dolly Moore recalled life as a student and volunteer at the school for forty years. The place had a rich character. "I can remember riding horseback to school, the outdoor bathroom, and the pot belly stove inside," she said decades later.

By 1924, the population of Seebe was down to forty people. In 1925, when G. H. Milligan moved to Seebe to work for Calgary Power, the community was just a whistle-stop. Other employees included superintendent F. J. Robertson, accountant Charles Robertson, and clerk Herb Biles. There were also eight operators for the two plants—on twelve-hour shifts, six days a week—and three maintenance men.

"Leapin' Lena," a Model A, was used to transport Calgary Power staff around the Seebe area, c. 1937. TRANSALTA COLLECTION

In 1929, the population of sixty included a schoolteacher, pastor, civil engineer, carpenter, merchant, storekeeper, and postmaster. That year Calgary Power sent electricity from the dams at Seebe to southern Alberta customers over three high-tension, 55,000-volt lines. Power from this facility also went to the Canada Cement Company plant at Exshaw. As of 1929, Calgary Power was supplying ninety-eight communities, including Morley, Cochrane, Okotoks, and High River, with power over 1,600 kilometres of transmission lines.

In 1937, an Alberta Women's Institute Girls' Club met regularly at Seebe. The Chickadees met twice a month, with seven girls on average at each meeting. Activities included a corn roast after one of their meetings, and they took in an interesting talk on "Pride in Personal Appearance." Knitting and crocheting were regular activities at their meetings, and they also held teas "at which they entertained their mothers." They hiked down to the river in April with a lunch, "the Club supplying the pop."

In 1941, the company started handing out employee numbers. Jal Abelseth of the Seebe plant was number 00001.

After World War II, the company completed its Barrier Generation Station on the Kananaskis River. It was the first remotely controlled hydroelectric facility in Alberta, and one of the first on the continent. Calgary Power automated all its facilities and built a central control centre in Seebe in 1951 to manage them all. It worked well until 1985 when a new system control centre in Calgary opened.

The Seebe School was closed in 1996, the last of approximately five thousand one-room schools across Alberta, dating back to the first in 1864 at Victoria Settlement on the North Saskatchewan River, east of Edmonton.

On Tuesday, 31 August 2004, the townsite of Seebe officially closed. Only memories remain.

Seebe in 1961. TRANSALTA COLLECTION

Calgary Power accounts remained relatively stable throughout the early 1930s, in spite of the Depression. The company's innovative program of installing electric motors in grain elevators across the province provided an important additional customer base, and Gaherty continued with his vigorous campaign to sign up new customers across Alberta. "The Calgary Power Company, greatly interested in elevator electrification as a load builder," reported the company newsletter, *The Relay*, "in 1933 decided all objections could be removed if an organized installation programme with all the advantages of standardization, mass production, and installation was undertaken, and commenced negotiations with the elevator operators and the electrical manufacturers…" In all, the company installed units in two hundred elevators in one hundred locations in June 1933. By 1937, Calgary Power had electrified 421 elevators for twenty companies, "representing almost complete saturation for grain centers where electric power is available, with the assurance that other elevators will be electrified as the company's lines are extended." *The Relay* continued, "Great things have small beginnings," noting that as a result of innovative people applying their "combined effort" to the task, over four thousand kilowatts of connection were added to the system by providing this standardized service to Alberta's grain elevators.

During the 1930s, the company developed a highly skilled team to design, construct, and operate new hydro facilities. Engineers from Calgary Power and Montreal Engineering worked together, including taking work projects in the other company's office in Calgary or Montreal. Their experiences encouraged them to be innovative and adapt their skills to local conditions. Where it was difficult to get heavy pipe into the higher regions of Kananaskis Country, for example, they fashioned lumber on site and created wooden penstocks to carry the water. And the engineers were also on site, learning firsthand how to estimate the runoff from the snow pack in the mountains.

Pulling Together

Staff morale was a major challenge during the 1930s. The destitute and unemployed sought work at the company gates, and single men—who did not qualify for relief—knocked on doors asking for food. In a bold move that was to pay great dividends, Gaherty convinced staff in 1936 to accept a 10 percent cut in salary in return for a commitment to avoid layoffs. Though the company rigorously cut operational costs, staff knew their jobs were safe. The policy boosted morale, built loyalty, and ensured the continuity of a skilled and experienced workforce, which the company needed in the long run. "Within three months after I bought the house, we got a 10 percent cut in salary," recalled Fred Buchan. "It was a hell of a lot nicer to get a 10 percent cut in salary than being laid off for lack of work. It wasn't too long before we got the 10 percent back."

In 1935, the company began the process of preparing for better times by hiring new engineers. The real prosperity, however, was still some years away, so Calgary Power put the recent grads to work scrubbing floors and digging post holes, all of which Gaherty claimed was a good way to learn the business "from the ground up."

While Calgary Power took care of its staff, the favour was returned many times over. One example took place in 1936. Every utility deals with risks to life and to property, and fire was a real concern for a company that relied on exposed transmission lines. In mid-November, a wind-driven, raging grass fire destroyed the #1 and #3 transmission lines from the hydro plants to Calgary. All available Calgary Power employees poured out of the city to help. Accountants dug holes for new poles, and salespeople strung out wire for the linemen.

In spite of the staff's heroic efforts, the raging firestorm won out and severed the power flow early in the morning. Cut off from much of its hydro power, Calgary had to rely on a limited supply of electricity from the #2 transmission line from the Ghost Plant, the #13 line from Lethbridge, and the connection to the East Kootenay Power supply in British Columbia. At about 9:22 A.M., "the hurricane [grass fire] reached our #2 line, a few miles west of Calgary," reported *The Relay*, "and broke down over 20 structures." The company brought the Victoria Park steam plant into service at 10:45 A.M. The severe winds then damaged line #13, and for a time the entire city relied on power from the steam plant. At 6 P.M., line #13 was back in service, and at 10:30 P.M., line #2 was working again. "Fortunately, except for Calgary, our service was practically unaffected throughout the day," stated *The Relay*. The City of Calgary lost more than one hundred poles that day: though busy with its own repairs, its "Electric Light Department assisted us greatly with men and equipment in repairing our lines feeding the city."

The fire consumed about 544 square kilometres of land in the Cochrane area, destroying livestock, buildings, and feed. Rain fell the next day, 20 November, and according to the *High River Times*, the precipitation killed the fire: "The fire and wind also crippled the Calgary Power service all day Thursday. The power line right-of-way around Seebe was a scene of desolation, with poles blown down and wires entangled. But by about ten o'clock Thursday night electric power was brought up to strength, streetcars again ran in Calgary and normalcy was restored."

According to Gaherty's 1936 year-end letter to employees in *The Relay*, Calgary Power staff had risen to the challenge of the fire "regardless of whether or not the work came within the scope of their duties." Poor crops, operating difficulties, and the economy all negatively affected the company's success. "The excellent work done by our employees has gone a long way to offset these adverse factors."

Even more serious were mechanical or electrical fires in generating stations. In September 1940, a fire broke out at the Horseshoe Falls power plant after a failure in one of the large transformers. A whole section of the building went up in flames. This forced both the Horseshoe Falls and the Kananaskis plants to suspend power generation, triggering a supply crisis for wartime Calgary. Once again, the call went out and all employees in the area rushed to clear away the debris, install a temporary transformer and switch, and get the power moving again. But, given wartime supply conditions, it was some months before a new transformer was installed and things were back to normal.

Not all was hardship during those trying years. The company continued a program—no one is quite sure when it started—of giving turkeys to employees at Christmas. In 1938, *The Relay* amalgamated all the notes of thanks it received into a poem, which read:

continued on page 60

Fire disaster at Horseshoe Falls, 1940. TRANSALTA COLLECTION

The *Relay* Story

Communication is of utmost importance to any group, no matter how small or large. And so, in the middle of the Dirty Thirties, with less than 150 employees on the payroll, Calgary Power created a Publication Committee. Its mandate was to publish a company magazine. From 185 suggestions, the name *The Relay* won out, and in the summer of 1936, company general manager G. H. Thompson introduced the first issue. Intended as a monthly publication, it began as a newsletter.

In the first issue, Herb Biles of the company reported that Calgary Power had connected 170 refrigerators to the power lines between January and May 1936, for example, compared with just 111 for all of 1935. "We hope the heat wave will keep on coming," reported *The Relay*.

Off and on until the 1990s, the little publication grew and kept the Calgary Power family up to date on recent events, most of which were serious and related to company business. But not always.

"Now a word about grizzlies," wrote W. R. Wolley-Dod, otherwise known as Pocaterra Pete. "The grizzly is one of the shyest animals that roam the hills and rather than encounter men will do his best to get out of his way…" He once scared off a grizzly, only six paces away, with a most unusual strategy. "I spoke nicely to him but he either didn't hear or didn't understand the language so I started lighting matches." The bear ran away.

But *The Relay* also explained the company's business to everyone in house. Country grain elevator electrification, the installation of electricity at the Turner Valley Gas Plant, the operations of Calgary Power's sister company: Montreal Engineering Co. Ltd., rural electrification, construction projects—dozens of topics in all. It included short biographies of senior staff. Bowling scores, births, illnesses, and social events all graced its pages.

Golden Jubilee Relay *cover, 1955.*
TRANSALTA COLLECTION

Louie the Lightning Bug and "Sir Care a Lot" mascots with Lou Sikora during the United Way campaign, 1989. TRANSALTA COLLECTION

In January 1942, Calgary Power "considered it expedient to suspend the publication of The Relay until conditions are more favorable," due to shortages of staff and materials during the war. In 1961, general manager Fred T. Gale used The Relay to provide a fifty-year synopsis of the company's accomplishments. Other articles in 1961 and 1962 provided a detailed history of Calgary Power's operations.

Vicki Slater, a wife and mother of two teenagers, became the company's first female meter "man" in 1976. "I'm not a women's libber, but I've always felt a person should pursue any career they were capable of, regardless of their sex," she said. "My co-workers treat me like one of the guys but not like a man. I'm really enjoying it and looking forward to the future."

In the late 1980s, The Relay highlighted the company's commitment to the environment with its support for Arbour Day. Though the company had to cut a lot of trees and brush away from its lines, it also sponsored tree planting programs in schools in 1989. Grade 3 kids in thirty schools planted lodgepole pines as part of a school program. "I hope my tree grows to be big and strong," wrote Steven, a participant in the Arbour Day program from the Percy Pegler Elementary School in Okotoks, Alberta.

Not dodging controversy, The Relay addressed the global warming issue in 1990, as president Ken McCready led initiatives to tackle the challenge and contribute to the solutions. The editor of the magazine noted: "Although the jury is still out on whether global warming is happening, TransAlta is taking action now."

The Relay ceased publication in the mid-1990s. In 2009, TransAlta began publishing a quarterly field newsletter, The Power Source, in print form and as an online publication. It continues the tradition of The Relay with its mix of news, employee and facility profiles, milestones, successes, and other relevant information, and keeps more than 2,400 employees informed on the continuing developments in the company.

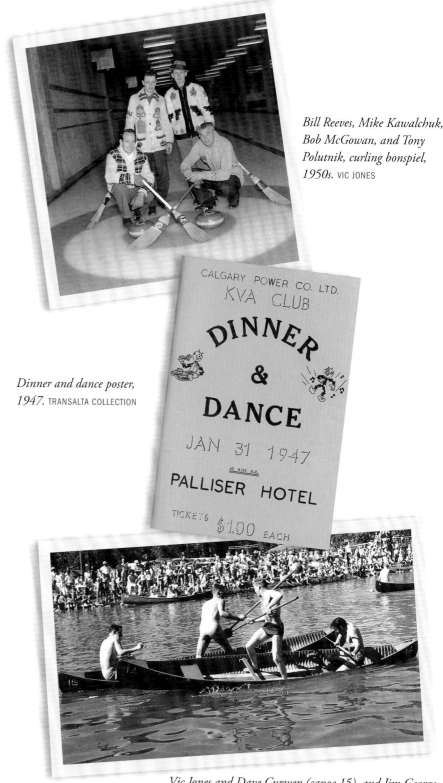

Bill Reeves, Mike Kawalchuk, Bob McGowan, and Tony Polutnik, curling bonspiel, 1950s. VIC JONES

Dinner and dance poster, 1947. TRANSALTA COLLECTION

Vic Jones and Dave Curwen (canoe 15), and Jim George and Tamo Takanaga (canoe 14), canoe jousting at a company picnic, Bowness Park, 1950s. VIC JONES

To the Management of Calgary Power
We extend our thanks at this gladsome hour
The thanks of the you's, and the thanks of the we's
For those big, fat, Christmas Tur-keys.

Calgary Power also organized special events for its employees. Picnics were popular, and included games and sporting events. The riverside locations of the power plants and a chance to get out of the city were attractive features for employees and their families, most of whom did not own cars. In 1937, the company picnic at Ghost Dam west of Calgary included boat rides, baseball, tug of war, swimming, kids' races, and refreshments for 136 company employees and dependents.

Calgary Power and "Funny Money"

Killam and Gaherty paid careful attention to provincial politics and strove to maintain a good working relationship with the government of the day. In 1921, the United Farmers of Alberta (UFA) defeated the Liberal government. Some UFA members supported rural electrification and government ownership of electrical utilities. But Premier John Edward Brownlee, R. B. Bennett's former law partner, was reluctant to consider any such heavy government expenditure.

By 1932, Alberta was virtually bankrupt and could borrow only from the federal government. As economic conditions worsened, protest movements emerged on the right and the left. That same year, in Calgary, various socialist, labour, and progressive groups founded the Co-operative Commonwealth Federation (CCF). They argued that the capitalist system had caused the Depression, was exploiting workers, and had to be replaced through co-ops and other forms of industrial organization. Calgary Power management knew that electricity had always been a favourite target for public ownership.

In this era of drought, destitution, and unemployment, the UFA government appeared impotent and indecisive. In 1935, Alberta voters turned their backs on the UFA and handed a massive majority to a religious movement espousing a dubious economic creed called Social Credit. "Bible Bill" Aberhart turned his evangelical radio program, "Back to the Bible Hour," into a formidable political tool, converting to his views the desperate voters of a poor province through the powerful new medium of radio. He created the Social Credit party, a grassroots political party that held power in Alberta for thirty-six years, until Peter Lougheed's Conservatives swept into office in 1971 using another new medium—television.

No one knew what to expect from Social Credit and the curious economic theory that every citizen deserved to benefit from the "social credit" inherent in the wealth and resources of the province. The SoCreds launched an attack on the Canadian banking system, currency, and financial practices. When prohibited from creating an

Turner Valley oil well, c. 1930. GLENBOW ARCHIVES NA-4058-1

Alberta bank, it formed the Alberta Treasury Branch—a near-bank. For many Albertans, it was a financial alternative to the "eastern" banks, disliked for their predatory lending practices and confiscation of property for unpaid loans—realities of the Great Depression.

And then, in April 1936, the SoCred government defaulted on its bond payments. The federal government grew alarmed with the situation in Alberta, especially with the province's challenge to Canada's centralized banks. Such a challenge threatened the unity and integrity of the Canadian financial system and Canada's international reputation. The federal courts disallowed, on constitutional grounds, many of the key pieces of provincial financial legislation and the more radical aspects of the Aberhart platform. With time, the SoCred policies mellowed and a more pragmatic premier replaced the idealistic radical. Social Credit politicians learned to work with investors, entrepreneurs, and capital from outside the province during the boom that followed the discovery of oil at Leduc in 1947.

But back in 1936, a bit of good luck visited the province when an independent oil company discovered crude oil at the Turner Valley oilfield southwest of Calgary. Small discoveries in 1914 and 1924 had attracted the attention of local investors, including R. B. Bennett and Sir James Lougheed. Major oil companies, including Imperial Oil of Toronto, had invested heavily in Alberta's first commercial oilfield in the 1920s.

But the Turner Valley Royalties No. 1 well in 1936 was a true gusher. As a result, Turner Valley production jumped to over 2 million barrels of oil in 1936. In 1937, it produced more than the people of Alberta could consume. More than 6 million barrels spewed out of the ground in 1938, prompting Standard Oil of California and Texaco to move branches of their companies to Alberta. Production grew to 8 million barrels in 1940, and then to nearly 10 million in 1942 as the field gave of its wealth for the war effort during World War II. Turner Valley produced about one-quarter of all the oil consumed in Canada in the early 1940s. More big oil companies flocked to Calgary, with Gulf Oil and Shell of California opening offices in the early 1940s.

The new markets sucked up some of the excess electrical capacity that had remained unused through most of the 1930s. With an optimistic eye to the future, *The Relay* predicted in 1938: "Turner Valley is merely one phase in the picture and other fields will be discovered that will startle not only Canada but the world." The Turner Valley oilfield's massive expansion in the late 1930s allowed Calgary Power to develop valuable experience that it would put into practice when the oil boom hit in the late 1940s and 1950s.

Reaching Out to Albertans

As economic conditions slowly improved in the late 1930s, Calgary Power worked hard to encourage its customer base, which was still limited, to use more electricity. In 1932, the company created its Home Service Department "to help our employees and customers make their homes more satisfying and enjoyable places to live"—using electricity, of course. Newspaper advertising campaigns in

Calgary Power advertisements, 1936.
TRANSALTA COLLECTION

Calgary Power's "Modern All-Electric Kitchen" won the Mercantile Section of the Calgary Stampede Parade in 1937. It then went on tour around the province.
TRANSALTA COLLECTION

1936, for example, suggested that sparing use of electricity was not good economy. A washing machine, a vacuum cleaner, a refrigerator, a reading lamp, and a radio could add greatly to one's "comfort and enjoyment" of life. And who could argue with the logic that "the more Electricity you use, the less it costs?"

In 1937, Calgary Power dispatched a travelling trailer-based "Modern All-Electric Kitchen" display with a six-person crew. The kitchen included a "range" (an electric stove), refrigerator, water heater, ventilation fan, washer, ironer, food mixer, coffee maker, radio, and a compressor for hot and cold running water. Home economists Audrey Dean, Marianne Pearson, Shirley Scott, and Flora Williams were decked out in immaculate white uniforms and charged with informing customers of the wide range of new "labour-saving devices" that could be plugged into the electrical outlet. At night the four women bunked in the display, which served as a dormitory on wheels, keeping warm in sleeping bags on air mattresses—one of

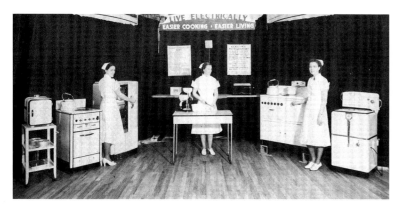

Flora Williams, Marianne Pearson, and Shirley Scott, Calgary Power Home Service department, 1930s. TRANSALTA COLLECTION

Supporting the War Effort

In September 1939, Nazi Germany invaded Poland, precipitating World War II. Once again Calgary Power staff were encouraged to enlist and promised jobs on their return. Fellow employees sent packages with cigarettes and other treats to those overseas.

Gaherty made it clear that Calgary Power was in the business to defeat Hitler and free Europe. Calgary Power's founder, Max Aitken—now Lord Beaverbrook—was Churchill's minister of Aircraft Production, and he appealed to Canadians to deliver as never before. Calgarians went to this war with no false sense of enthusiasm. Given their earlier experience, they were less innocent than in 1914, knowing what war would entail.

As during the previous war, employees had to deal with long hours and broader job responsibilities to cover for the men and women who enlisted. In a 1941 speech, Gaherty warned that labour shortages were a barrier to the effective delivery of electricity for the war effort. The war provided a new opportunity for female employees to expand their roles and break down gender barriers.

The supply of electricity in Alberta quickly became a key ingredient in the war effort, especially after the shock of the defeat of France in June 1940. Until the attack on Pearl Harbor in December 1941, the U.S. was not a factor in the Allied cause. For two years, Britain and the Commonwealth stood alone in Europe against the Axis Powers. While Canadian troops were important, the food, raw materials, and military supplies that Canada could supply were equally valuable to beleaguered Britain, which needed all the grain, pork, beef, and other foods that Albertans could ship across the ocean—and cold storage plants required electricity. Ottawa also financed new Alberta factories to produce munitions and explosives for the war. The federal government was determined to ensure that these factories had adequate electric power supplies.

which leaked air. They toured the province visiting fairs and rodeos and riding in parades—not just selling new appliances, but working to professionalize the roles of females in the home.

"The Trailer," as it was called, won first prize in the Mercantile Section in the Calgary Exhibition and Stampede Parade, "beating out our old friend the Gas Company and other entries," reported *The Relay*. The rivalry between electricity and gas had a long history, and after the parade, the kitchen display went to the Stampede Grounds "where it was the subject of much favorable comment by people attending the exhibition."

In another example of outreach to the community, Calgary Power donated the G. A. Gaherty Trophy to the Calgary Stampede in 1938. The trophy was sculpted by Charles A. Beil of Banff, renowned cowboy artist and associate of artist Charles Russell. It depicts a rider hitting the ground under the bronc, a common experience for those who competed in the North American Bucking Horse Riding Contest. Each year until 1942, a trophy was given to the winner of the saddle bronc event in the name of the president of Calgary Power.

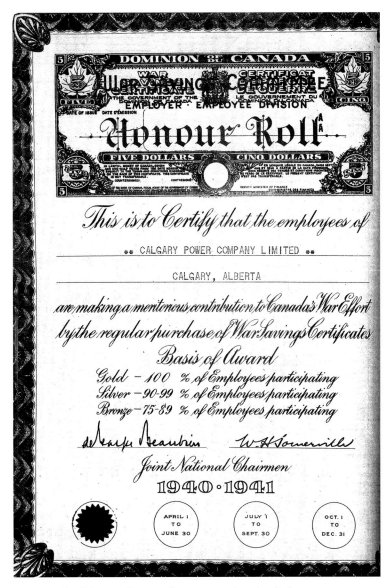

An example of the certificate employees received in thanks for supporting World War II efforts. TRANSALTA COLLECTION

The power supply issue came to a head in 1941 when the Alberta Nitrogen Company wanted to construct new facilities in Calgary and Edmonton to produce ammonia for the explosive, TNT. Calgary Power's reserve capacity from the Ghost Dam was being consumed, but there were potential new hydro sites at Spray Lakes and Lake Minnewanka that could help meet this demand. The latter was the preferred site given its proximity to the CPR lines and the Banff highway, but Parks had already rejected the project. However, the alternative site at Spray Lakes had access problems that would slow construction.

Discussions between Calgary Power and Ottawa officials focused on Lake Minnewanka, with its large storage and potential for power generation. Parks used the tools available to it to delay, and hopefully to stop, the project. It argued that no approvals should be granted until parliament approved changes to the 1930 National Parks Act. C. D. Howe, the federal minister of munitions and supply in the Mackenzie King cabinet, overrode the objections of the Banff park officials, and cabinet approved the work at Lake Minnewanka. Matters relating to the war effort had priority.

In response to the discussions in Ottawa, Calgary Power applied to raise the level of Lake Minnewanka a further twenty-five metres—the company had earlier increased it by sixteen metres. This significantly changed the size and nature of the lake and its aquatic ecosystem. Minnewanka doubled in length to 28 kilometres, and the surface area increased by 50 percent. Large areas of the shoreline flats were flooded, as were cottages and the inn.

Calgary Power also temporarily diverted part of the flow from the Ghost River into the east end of the lake to help fill the reservoir. This water, which normally would only have generated power once as it went through the Ghost plant, ended up flowing through four generating plants: Cascade (Minnewanka), Kananaskis, Horseshoe, and Ghost. The enhanced storage capacity provided by Lake Minnewanka allowed Calgary Power to deliver greater peak power, even in winter.

This was the system that Aitken had envisaged after the company's early miscalculation of winter flow rates.

The company built a huge curved landfill dam in 1941 and a canal to deliver the water to penstocks that carried the flow down to the Cascade power plant beside today's TransCanada Highway, just east of Banff. A new transmission line linked it into the power grid at Kananaskis Falls.

On Guard

The Battle of Britain in 1940 brought a much more serious tone to the war for Calgarians. A letter to staff at the time warned all employees to be on their guard and to report any suspicious activities: "When Canada is involved in war, the possibility of sabotage will exist, therefore it is necessary for every man employed by a public utility to be on guard."

Security at the plants increased when two prisoner-of-war camps were established in the area. The Seebe Internment Camp No. 130 held 650 prisoners, among them conscientious objectors, enemy seamen, and German officers captured in North Africa. A smaller camp near the Barrier Dam site on the Kananaskis River also held prisoners of war from 1939 to 1945. Militia units protected the power supply, but Calgary Power employees also carried weapons.

We had two prisoner of war camps near Seebe … and people were a bit nervous having German prisoners so close to their

Armed personnel carriers protecting hydro plant during World War II. Kananaskis Falls hydro plant. TRANSALTA COLLECTION

Ernie McLeod and Ken Millar armed with guns, in the Ghost control room, 1939. TRANSALTA COLLECTION

DEPRESSION, WAR, AND SURVIVAL

homes. The operators and floormen at the plants had .38 revolvers in the control rooms at the plants and we had a 24-hour army guard at the north end of the Kananaskis Dam.
—Ted Robley Jr., Calgary Power employee

This certainly made employees feel part of the war effort, but it also created fears for those working lonely shifts on dark nights with their government-issued guns at the ready. Fortunately, there was no sabotage, but the jokes continued for several decades about the "firepower" of the Calgary Power "army," which brought back many trophies during hunting season.

Power for the Farmers

The postwar period brought prosperity, turbulence, and change to Calgary Power. Rural electrification had long been delayed in Alberta, and farmers wanted to enjoy the convenience electricity could bring. Most villages and towns, and all Alberta cities, had boasted electric lights for a generation or two, but farms were still without power after the war.

There were exceptions. During the booming 1920s, Calgary Power had signed contracts with twenty-four farms in 1927 near Brant, southeast of Calgary. But the company had to dismantle part of this distribution system during the 1930s when fewer than half the farmers were able to pay their bills.

Many farm and labour groups blamed Calgary Power for the slow pace of development, and demands surfaced once again for the government to take over public ownership of the utility. Some of the groups noted that a government-owned utility, the Manitoba Power Commission, had embarked on a program in 1945 to link farms to the grid—a system that by 1953 had delivered power to thirty-five thousand of Manitoba's fifty thousand farms. On the other hand, Saskatchewan had lagged behind. Its government-owned utility, formed in 1929 as the Saskatchewan Power Commission, was unable to provide power to farmers due to large distances and a sparse population. And it had no access to hydroelectric power. By 1949, only fifteen hundred farms had power, the same year the Saskatchewan provincial legislature passed the Rural Electrification Act. As of 1952, only four thousand farms had electricity.

Alberta's situation was not much different, and it became a hot issue in the 1944 provincial election. A 1944 study by the Alberta Post-War Reconstruction Committee concluded that "a well planned scheme of rural electrification" would modernize life on the farm. But it would be costly. "Not including wiring of buildings and purchase of machinery and appliances, [it] would range from about $500 to $600 per farm [or $6,400 to $7,600 in 2011 currency] … The Committee is of the opinion that the capital cost of rural electrification should not all be passed on to the consumer in the form of higher service rates, but should be borne in part by the state."

Another estimate put the cost to deliver electricity to each farm at $1,100 ($14,000 in 2011 dollars), with the full expense falling on the shoulders of the farmer. The SoCred government had considered funding the system, but it balked at the massive $200 million cost, an investment twice the level of the provincial debt.

After Premier Aberhart died in 1943, the quiet-spoken and shrewd Ernest Manning took control of the province. The new premier was a devout Christian, and conservative in his approach to the role of the state in society. He believed his Social Credit government should monitor and set guidelines for the economy, but should not own the means of production. Yet everyone, including Calgary Power president Geoff Gaherty, knew that Manning's core support in rural areas might force the Social Credit government into more radical action.

When Manning was accused of being too cozy with Gaherty and Calgary Power, the premier responded in 1944 by creating the Alberta Power Commission to come up with a solution to the problem

of rural electrification. Commission member Herbert Cottingham concluded that public ownership of the electrical utilities was the cheapest solution. A Crown corporation, he suggested, could be "a benevolent one and not for profit," and "the benefit of the natural resources should be shared by all."

In 1945, Calgary Power and Canadian Utilities—the other main electricity provider in Alberta—worked with the Western Section of the Canadian Society of Agricultural Engineering on a farm electrification experiment. Its goals were to "study various types of electrical transmission in rural areas with maximum economy in order to make it feasible to build lines … and to study the possibilities for use of electrical power in the farm yard and farm home, in order to build up a load that would warrant the cost of lines and equipment."

Calgary Power chose the mixed farming region around Olds, 90 kilometres north of Calgary, and Canadian Utilities provided electricity for dairy farms 100 kilometres northeast of Calgary (near Swalwell), for thirty farms 103 kilometres east of Edmonton (around Vegreville), and for farms 95 kilometres southeast of Prince Albert, Saskatchewan (around Melfort).

A young engineer, E. Bruce Martin, monitored the project for the federal government. His desk sat in the corner of the Calgary Power office in Olds, and Fred Gale and other company employees helped him survey the 112 farms in the test area. For two years, Martin watched with interest as farmers and their families responded "to the magic of the power in the little insulated wires, and worked with ingenuity to adapt hand-operated equipment to an electric motor." Martin considered the experiments a success because the companies watched as "the farmers learned to use electricity to the maximum" and were able to learn firsthand the best way to deliver electricity efficiently to a rural area.

In July 1947, Manning exhorted Calgary Power and Canadian Utilities "to proceed at once to put into effect a rural electrification

Woman using rotary arm iron, c. 1940s. TRANSALTA COLLECTION

program for Alberta to provide such service to not less than 21,500 farms within the areas presently served by their main transmission lines, to be fully accomplished within a maximum period of ten years." At that time, Calgary Power had just four hundred farms hooked into its system. While many rural voters endorsed the plan, urban residents feared rural expansion costs would increase their rates, because the premier's plan included uniform electricity rates across the province.

Pressured by the rural members in his caucus, Premier Manning announced a modest plan to add about two thousand farms a year to the electricity system. In a letter to Calgary Power and Canadian Utilities in early 1948, he firmly stated his objective. "The Government of Alberta would prefer to have the program, herein outlined, implemented by the present companies, but the development of the province has reached the stage where a comprehensive program of rural electrification cannot longer be delayed." As an added incentive, the province promised a 50 percent refund of the corporate income taxes paid to Ottawa—a rebate being provided to Alberta because most other utilities in Canada were owned by government and therefore not taxed. Gaherty said the refund was inadequate: there was no return on the capital invested or recognition of business risks. And he suggested that, since rural electrification was a provincial priority, the Alberta government pay 50 percent of the costs. Manning's proposal was attracting little enthusiasm from industry.

The premier embarked on a bold political gamble. He called a provincial election in August 1948 and put the debate over public ownership of the electrical utilities to the voters to decide in a plebiscite. This move allowed Manning to defuse the strongest plank in the CCF platform and separate the Social Credit party from the contentious issue of rural electrification. The CCF leader, Elmer Roper, had promised the immediate takeover of the electrical utilities and "country power" for all, at no cost to the farmers. Roper considered it unfair to allow urban voters to cast ballots in the plebiscite as they had "no direct interest in getting electricity into the farm homes." But Manning replied that the CCF's "idea of democracy is to shove things down the public's throat." With any luck, the plebiscite would favour private ownership of the electrical utilities in the province, and the pressure of the voters would convince the privately owned utilities to step up to the task of providing power to rural areas.

Calgary Power certainly felt the pressure. It seemed its very existence was on the line. Many Calgary Power employees took an active part in the election and the plebiscite, because they felt the premier was not defending free enterprise and their jobs strongly enough. The first set of announced results was encouraging for the company, with a nine-thousand majority in favour of the status quo. Calgary Power employees were ecstatic.

The Plebiscite

During a challenge to its very existence as a private company, Calgary Power encouraged its employees to get out and vote in the 1948 provincial election. Appearing along with the vote for MLAs was a plebiscite asking Albertans if they wanted Calgary Power turned into a government-owned public utility. Voters narrowly defeated the proposal.

"The reason why I remember this," recalled Gordon McKenzie many years later, "is because that was when my first son was born—August 8, 1948. My wife was in the hospital at the time this plebiscite was being held. I remember Happy [Hansen] coming up to the office to anyone who was around at 3:00 in the afternoon, to make sure we all voted on the plebiscite, and asked if our wives voted, and our relatives. My wife hadn't voted—she was in the hospital—so he gave me the keys to his car to go and get her to take her over there to vote, which I did."

ELECTRIFICATION PLEBISCITE

Do you favour the generation and distribution of electricity being continued by the Power Companies at present?

OR

Do you favour the generation and distribution of electricity being made a publicly-owned utility administered by the Alberta Government Power Commission?

Mark the figure "1" next to your choice.

The plebiscite question put to the Alberta public in the 1948 provincial election.

Manning made his support evident: "I am particularly pleased to note the emphatic repudiation of the socialist threat to our free democratic way of life." However, there were problems with the count on election night, and it took weeks for the official results to be released. The majority shrank to 703, and later to just 151, of the 279,831 ballots cast in the plebiscite. The results from the Vermilion district were particularly in dispute, and eventually 57 percent of the votes in that riding went to public power.

In the end, the way the vote split correlated with whether or not the voters already had electric service—and not with any political leaning. Voters in communities already serviced by Calgary Power and Canadian Utilities voted for the status quo. For example, Calgary and the towns along the Bow River voted 65 to 70 percent in favour of private power. But in Edmonton it was close—voters favoured the status quo by only 51 per cent.

Farther north, in areas with little or no electricity, profoundly conservative voters who cast their ballots for Manning's Social Credit party on the election ballot, turned around on the plebiscite vote and supported the idea of the government taking ownership of the electrical utilities. Some areas, such as Redwater (77 percent) and Willington (72 percent) delivered strong majorities for the provincial takeover. Others such as Edson, Athabasca, and Pembina were close behind. The public perception in these districts was that the province would provide rural electrification more cheaply and quickly.

Removing Calgary, Edmonton, and Red Deer from the vote, the plebiscite results were 104,235 in favour of public power versus 88,532 for power from the private utilities. Including the votes of Calgary, Edmonton, and Red Deer, the split was 139,840 in favour of public power, but 139,991 in favour of the private utilities. Thus, even though rural voters, who were directly affected, cast their ballots in favour of the government taking over the utilities, the ridings with installed electricity carried the day.

It was a close call for the privately owned electrical utilities in Alberta, but not for Manning. The Social Credit party took 55.6 percent of the votes, and fifty-one of fifty-seven seats in the general election, and the premier had a strong bargaining tool when he negotiated the rural electrification plan with the investor-owned utilities. Calgary Power agreed to Manning's program, took a deep breath, and quickly mobilized to expand its network to meet the needs of rural residents.

Earlier in the year, on 3 February 1948, Calgary Power had replaced its Rural Department with a not-for-profit subsidiary called Farm Electric Services Limited. It expanded its service to farmers and built transmission lines out into the prairies, at cost, to local co-operatives. Partly financed by the province, the Rural Electrical Associations (REAs) were based on an idea that had come out of experiments by the Roosevelt Administration in the U.S. Many REAs were operational before the election; six sprang up in 1947 and thirty-four more in 1948. Calgary Power officials were out everywhere explaining the system and helping locals organize their own associations. Neighbours sold the idea to neighbours. Rural electrification eventually met all of Manning's goals, and then some. The premier had set as a target the goal of connecting 21,500 farms to the grid by 1956, and when that year arrived Calgary Power was already serving—directly or indirectly—30,000 farms. By 1961, 87 percent of Alberta farmers had power.

The province played a key role in the development of the system by guaranteeing the loans for the REAs and covering 50 percent of their costs. The Alberta Power Commission regulated the associations, and their loans were managed through the Co-operative Association branch in Edmonton. The Public Utilities Board set the rates to ensure they were fair to the local co-operatives. And the government allowed farmers a 10 percent deduction for their costs against their income tax for ten years. As the momentum built, farmers saw electricity as a cost-effective, labour-saving tool that quickly

continued on page 72

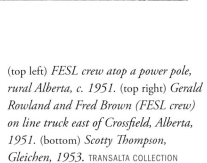

(top left) *FESL crew atop a power pole, rural Alberta, c. 1951.* (top right) *Gerald Rowland and Fred Brown (FESL crew) on line truck east of Crossfield, Alberta, 1951.* (bottom) *Scotty Thompson, Gleichen, 1953.* TRANSALTA COLLECTION

Turning on the Lights

At Leduc West, George Parkinson was working for the company the winter evening when it turned on the power for the rural residents of that area.

"From where we were, you could see the lights come on in various places. So we immediately starting going around to every service [at each farm], to make sure everything was all right. And every so often, we'd open up a meter box, and here's a mickey of whiskey sitting there, or a bottle of beer. And I'm glad there were no checkstops out that night, otherwise we'd never have passed them! Everybody was just so happy to see the power come on. They'd all worked and helped put this thing up, and it was quite a night."

At a farm northeast of Edmonton, Parkinson was able to see the kind of difference the installation of electricity made in a family's life: "As we hung the transformer, we would go into each house to make sure everything was all right ... I went over to the house, knocked on the door, and they let me in. The lights in the kitchen were on, nice, brightly lit and everything like that. The elderly lady, the grandmother, was sitting in her rocking chair, just rocking back and forth with a smile on her face. I looked up on the shelf by the side of the stove, and there were six or seven kerosene lamps up there. The real fancy ones, with the nice colours through them and so on. The old lady looked up and said, 'You know, young man, I have polished those lamps for the last fifty years, and thank the Lord I don't have to do it again.'

"Another occasion, when we energized the farm, it was mid-afternoon, I suppose. We hung the transformer and when all was ready, we closed the breaker. And there was the worst scream that came out of the house that you could imagine. We, here's the three of us, with our belts and [pole climbing] hooks on and everything, and we went tearing over to the house. We were sure it was somebody got electrocuted! Well, here's the lady of the house, standing in the middle of the kitchen, jumping up and down and screaming: 'The lights are on, the lights are on, the lights are on!' A little boy, about ten years old, was running around the house, flipping every switch in the whole place. And she was just so excited that the lights had come on."

—(*From 75th anniversary interviews*)

Calgary Power logo, 1940s.
TRANSALTA COLLECTION

demonstrated its value to all. It was also seen as a means of keeping sons and daughters on the land, slowing the exodus to the cities.

These were exciting and rewarding years for the employees at Farm Electric Services Limited. From 1948 to 1964, a social revolution swept across rural Alberta. Farmers installed electric lights, water pumps, and indoor plumbing, electric stoves and refrigerators, radios and milking machines. When an orange-and-black Calgary Power truck drove down a country road, everyone came running to help or supervise. An auger on the back of the truck excavated the hole, and dozens of hands lifted the poles and helped string the wires. Once the work was completed, everyone gathered for the ceremony of turning on the power. At the end of a long career with the company, Gordon McKenzie recalled the thrill of those times:

> I guess as far as satisfaction goes, I'd have to go back to the late 1940s and early 1950s, when [we installed] rural electrification… Believe it or not, [they] had always lived and worked without electric service. To get this to them, and the appreciation and what it meant to these people is pretty satisfying … No matter what time of year it was, January or July, it was New Year's Eve when the electricity came on.

The Leduc No. 1 discovery well, February 1947. TRANSALTA COLLECTION

72   POWERING GENERATIONS

The joys of modern life had finally come to rural Alberta, and Calgary Power employees took great satisfaction in being part of this transformation.

Rural electrification associations were an important development in the Canadian Prairie Provinces. The REAs demonstrated that having government, companies, and local people working collaboratively could control costs. Decades later, in 2011, the Alberta Federation of Rural Electrification Associations proudly supports its purpose: "We are committed to promoting the economic welfare and value of our members by providing strong representation to government and industry stakeholders with 'one voice.'"

While rural electrification was gaining momentum, another major event helped transform the Alberta economy. On 13 February 1947, the Imperial Oil Leduc No. 1 oil well struck a new source of oil southwest of Edmonton. It and other major discoveries out in the prairies proved that Alberta's oil was not all locked up in structures near the mountains, like the ones at Turner Valley. Production from these new oilfields eventually allowed Alberta to drill thousands of oil and gas wells, build pipelines to the west coast, to California, to the American Midwest, and to central Canada. Calgary Power was well positioned to provide electricity to the booming oil patch, and by the mid-1950s its lines were connected to thousands of oil wells, pumping stations, refineries, and gas-processing plants. The oil patch provided cheap natural gas, for fuel, to the first Calgary Power thermal units as the company began expansion into new generating capacity that also took advantage of Alberta's other hydrocarbon bonanza—coal.

The Killam-Gaherty era started with a "trial by fire" of the utility company, what with the Great Depression, a world war, and a plebiscite that could have altered Calgary Power's history. Strong leadership, solid financial underpinning, exceptional technical skills, a commitment to employees, broadly based markets, and an aggressive relationship with provincial and local politicians helped the company overcome these challenges. Its on-going ties to its growing urban and rural customer base allowed the company to expand its operations through the 1940s.

In 1950, Killam and Gaherty could survey their accomplishments with pride. They had inherited a strong and successful company and had built on those foundations. Calgary Power, now almost fifty years old, was well positioned to expand its role as the dominant provider of electricity in Alberta just as the company entered an era of unprecedented provincial growth and prosperity during which electricity would play a leading role.

Crowd at opening day, Leduc No. 1 discovery well, 13 February 1947.
GLENBOW ARCHIVES IP-6F-16

> "The company has been innovative. It has been forthright in its conduct with the public, it has been a forerunner in power systems technology, and it has been far-sighted in its planning."
>
> BERT HOWARD, TRANSALTA PRESIDENT, 1986

CHAPTER FIVE

The Move to Thermal Power

The three decades following the discovery of oil at Leduc were years of transition and change for Calgary Power. Employees felt a strong sense of pride and accomplishment, having survived the dark days of the Depression and the difficult years of World War II. They looked forward to better times. The new oil fields going into production each year triggered rising expectations. Capital and expertise poured into the booming province, and the demand for electricity exploded.

As it entered the 1950s, Calgary Power owned 65 percent of the electrical generating capacity in Alberta. In the postwar years, the company experienced an unprecedented level of growth: its customer base doubled, income increased fourfold, plant generating capacity grew sixfold, and overall generation increased from 90 MW to 3000 MW. And it was during this period that Calgary Power made a decision that would define its future for many years.

On 12 May 1947, the company incorporated itself as a new entity, Calgary Power Ltd., taking over all the assets of Calgary Power Company, which, according to the annual report for that year, "since 1911 had been engaged in the production and sale of electricity in the Province of Alberta."

At this time, Calgary Power also moved its headquarters from Montreal to Calgary. But its pivotal links to Royal Securities and Montreal Engineering remained, allowing it unparalleled financial and technical support. Company president Geoffrey Gaherty continued to

Turbine at Wabamun, 1960s. TRANSALTA COLLECTION

reside in Montreal and provided engaged leadership and solid management for Calgary Power's operations until his retirement in 1960.

But the first order of business in 1950 was yet another challenge from the City of Calgary. Some members of council questioned the city's purchasing of electricity from Calgary Power. The contract with the city represented 25 percent of Calgary Power's sales, and Gaherty had no intention of losing the corporation's largest customer to a proposed city-owned generation plant. Construction of the Spray Lakes power plants, now underway, assumed the continuation of the city contract.

The City of Calgary's electricity department commissioned an independent consultant to investigate the alternatives. His report recommended that the city construct and operate a new 110 MW gas-fired power plant that would eliminate the need to renew the Calgary Power contract when it expired in 1953. The estimated cost of $12 million seemed reasonable, and the report attracted interest and support in the city.

Gaherty challenged the consultant's report and offered his own estimates for a gas-fired generating station, one that Calgary Power was prepared to build just beyond the city limits. Although Gaherty's cost estimates were 50 percent higher than the consultant's, he had strong arguments for swaying the city. Did the city really want to take on the risk of financing such a major project when Calgary Power was

Calgary Power head office on First Avenue and First Street SW, Calgary, c. 1950. TRANSALTA COLLECTION

willing to pay for the construction of the new plant out of its own financial resources? Calgary Power's hydro option, Gaherty argued, was much less expensive in the long run. All City council had to do to take advantage of the low-cost alternative was to agree to extend the existing contract for another ten years when it came due. In the end, council decided to accept Calgary Power's offer of assured cheap power. Aldermen voted to review the idea of investment in electrical generating by the city in 1955. As it turned out, when the time came,

Calgary Power logo, 1950s. TRANSALTA COLLECTION

the city's demand for power was growing so quickly that only Calgary Power's massive network could possibly keep up to it.

The oil discoveries around Edmonton created an enormous demand for electricity in the capital region. Calgary Power developed a special working arrangement with Imperial Oil to supply power and specialized services for oil and gas production facilities in the oil field and for a new refinery near the capital. There were other major customers: British American Oil, Canadian Oil, Texaco Canada, and Shell Oil Canada. Hundreds of smaller firms involved in seismic exploration, geological assessment, drilling and drilling supplies, pumps, derricks, pipes, pipelines, and refinery equipment were also hungry for electricity.

The boom called for rapid expansion of the distribution system and its transmission lines, often to isolated sites, and required late night and weekend work to install facilities on demand. But there were few complaints. Employees felt they were building the new Alberta. There were risks—if an oil field did not prove commercial, Calgary Power could end up with stranded assets. But that seldom happened. The company's customer base grew by leaps and bounds. This was a particularly satisfying achievement, as sales to oil companies took considerable effort to establish. Employee Gordon McKenzie recalled:

> The company had to struggle to sell electricity to oil companies for their field operations, because of the competition for business in the oil fields. Since the oil fields were developed primarily by American companies who, historically, no matter where they operate in the world, would burn their own by-product—natural gas [in order to make electricity] to do any pumping or whatever they had to do. Electric energy was completely foreign to them. It took a lot of time and work to make some inroads there and to meet that competition.

Wabamun steam generator

The general area designated (4) in the diagram represents the superheaters. In this section of the generator steam at boiling temperature (212° F.) is admitted, but as it passes through this superheater section (which is directly in the path of hot furnace gases) its temperature is raised to 900°. The other sections of the steam generator, not a part of this story but interesting to note, are: (1) burners; (2) water walls; (3) drum; (5) economiser; (6) air heaters

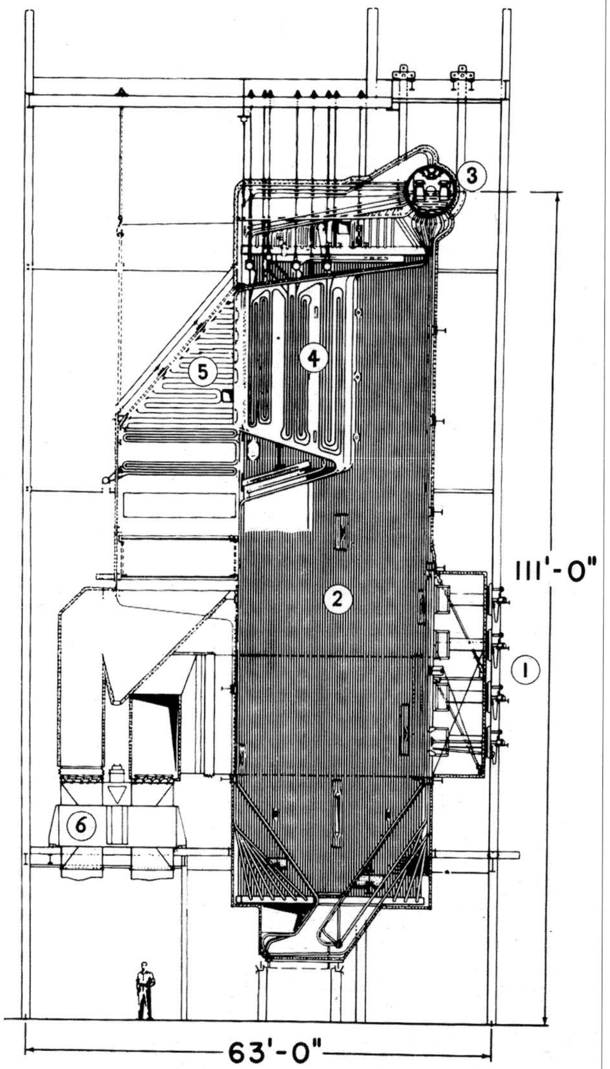

Diagram of steam generator, The Relay, *summer, 1960.* TRANSALTA COLLECTION

In this exciting new era, Calgary Power's biggest challenge was to increase generating capacity at the same pace as customer needs were growing. If the company moved too slowly, it risked opening the door to competitors. If it moved too quickly, it would suffer the economic drag of unused capacity. During the late 1940s and the 1950s, the company bought power from utilities in southeastern British Columbia, built three new hydro plants, expanded two others, commissioned its first thermal gas-fired unit, and substantially expanded transmission and distribution, including rural electrification. All of this activity forced Calgary Power to focus on decision making and project management as never before.

Project management emphasized a combination of speed, thoroughness, accuracy, and public accountability. Each step in the planning process required careful research and interdisciplinary analysis. Market need was a given during the post-WWII boom. Many fuel options presented themselves in resource-rich Alberta, each requiring careful cost analysis and projection into an uncertain future. Location choices were complicated, yet it was generally easier to move the power than the fuel. However, in the case of hydro options in the far reaches of northern Alberta, transmission costs to get electricity to market outbalanced the relatively low cost per KWh at source. As for design decisions, Calgary Power's choices were typically the best available within strict budgets.

Regulatory hurdles in those years were simple and straightforward, because the province did not require public consultation nor formal hearings. Government approvals were not complex during the 1950s and 1960s—industry, the public, and government were of one mind when it came to development—but these relationships grew more complex as public consultation became the norm in later decades. Construction cost inflation eventually became a concern, but Calgary Power, as the largest utility company in Alberta, had economies of scale that afforded it leverage over its contractors.

Project Management
1. Assessment of Market Need (How much new power?)
2. Fuel Options (hydro, gas, or coal)
3. Location (fuel source and market proximity)
4. Transmission (cost-effective delivery to market)
5. Design (engineering and cost)
6. Public Consultation (interaction with stakeholders)
7. Government Approvals (regulatory and political)
8. Construction (on time, on budget)
9. Financing (capital cost and availability)
10. Commissioning (integration and grid connection)

A completed project went through a final integration of all components to ensure that the system functioned properly. A new plant had to operate reliably, with maximum hours on line per year, and at the lowest possible cost. Technical excellence allowed for cost control. And, finally, the company needed to generate a 10-percent-plus profit.

Tapping the Bow

In the 1950s, Calgary Power launched an aggressive program of hydro expansion in southern Alberta to meet the demand for power. The Cascade power plant near Lake Minnewanka had been built during World War II, and the company had initiated several other projects in 1945. Barrier Lake Dam and reservoir, south of the TransCanada and Highway 40 junction, added 12 MW of new capacity. It took two years to complete and brought with it an important new innovation in 1951: remote control of the hydro plants. From the Seebe Control Centre on the Bow River, operators efficiently monitored and managed water flows at all the dams in the Bow River watershed. Patrol vehicles were supplied with mobile radios and were using VHF (very

continued on page 80

THE MOVE TO THERMAL POWER

Safety in the Workplace

"Since 1911 TransAlta has provided safe, reliable electric services to Albertans," the company's annual report boasted in 1990. Electricity can be a dangerous commodity to produce, distribute, and then use. Thus, safety has always been a concern for the company.

When the company was young and small, safety was taught through the mentoring process. In 1941, *The Relay* spread the message of the company's Safety First program:

SAFETY FIRST, is contagious, persist in practicing it and your fellow employees will be smitten.
SAFETY FIRST, when put in practice, forms habits of the right kind.
SAFETY FIRST, when persistently practiced, pays big dividends.
SAFETY FIRST, is as much a part of your regular duties as any other duty you have to perform.
SAFETY FIRST, your Company Officials not only believe in it, but are all out for it.
SAFETY FIRST, the success of which depends on the individual.
It is my wish that you and yours enjoy a "SAFER", and because of that, A HAPPY NEW YEAR.
– J. R. Jenkins

In the early 1950s, Calgary Power approached Hank Bradley, asking him to oversee the safety work. "Being the person who's done everything wrong once," Hank recalled years later, "I decided that the company was looking for someone of that nature. So I decided to take a try at their safety work."

Working as the company's safety supervisor "involved making trips around Alberta to the various locations of Calgary Power, inspecting tools and equipment, and keeping records of happenings throughout the year. And at the end of the year, holding safety meetings to review what had taken place. That way we could see if we could find the answers to what had happened and what could be done. This is how it kind of grew. My thought was that with the years between us all in the company, surely we could come up with a lot of right answers for how we could bring safety to our workmen.

"First aid is something everyone should not only know, but should be able to administer rapidly and systematically, as did Stu Thomas and Don Dowling at Violet Grove in 1956," said Hank. These Calgary Power employees rescued an unconscious man off a power pole after he made contact with a high voltage line. "A combination of fast thinking, pole-top resuscitation, and modern communications kept a serious accident from being a fatal one," reported *The Relay*. Taking advantage of a radio-telephone, they were able to call for an ambulance and keep William Hay of Wetaskiwin alive until help arrived. He fully recovered from his injuries.

"The first [safety] training schools were started in 1975," Hugh Cameron recalled. "We had three of them here in Calgary. I was on loan to the training school because we decided they should have First Aid training, and I was the First Aid training supervisor. We also thought they should have defensive driving because of the number of company vehicles out there. Initially, the program was set up as only a two-week program ... then it progressed to five weeks' training. Our first-year boys up at Red Deer get five weeks of training, which consists of defensive driving, First Aid, CPR, and pole-top rescue. Since 1975 when we started teaching First Aid training on the job, we have saved four of our own people—three linemen and one of our fellows at the plants. That's when it pays off. And the public—we've had just numerous cases."

By the mid-1980s, the company had safety incentive programs at its generating plants, and it created a safe driver award program in 1985 to encourage safety on the roads.

The Canadian Electrical Association tracks safety at operations where more than five hundred workers are employed. TransAlta had the lowest frequency rate of injury accidents throughout the late 1980s and the 1990s. In 1995, the company's generation business unit went for a full year without a lost-time accident, a success TransAlta attributed in part to a program that helped "employees identify and eliminate unsafe work practices."

In 1996, the company set as its goal an objective of "zero lost-time accidents" and worked toward this high standard by implementing new ways of analyzing the causes of incidents and reporting throughout the company.

Not content to strive for safety only at their own jobs, in 1998, TransAlta linemen volunteered their time to teach the basic rules of electrical safety to elementary school students.

In 2000, safety programs really paid off. Training programs, regular safety topic reviews, and constant improvements helped reduce the injury frequency rate and resulted in the company receiving the Canadian Electrical Association's President's Award of Excellence. In 2001, the injury frequency rate dropped 20 percent, due in part to the ten hours of safety training completed by each employee.

During 2002, the company officially launched the Target Zero program, aimed at eliminating lost-time injuries as well as medical-aid injuries. To that end, it increased safety training, held safety summits for supervisors and managers, and encouraged employees to report "near-miss" incidents as a way of eliminating hazards before injury occurs. From 2000 to 2003, the company's IFR—injury frequency rate—fell from 2.30 to 1.47, an improvement that pointed to a "cultural shift" in safety matters at TransAlta, according to Senior Safety Advisor John Cundall. Safety success continued through the decade and demanded constant attention. In 2006, a program was added to identify safety hazards. The safety rate for company employees fell as a result in 2007. In 2008, TransAlta created the annual President's Awards for Safety and recognized, by region, the employees and contractors who worked to create safe work environments. In 2009, the injury frequency rate fell below one injury for every 200,000 hours—significantly below the industry average—for the first time ever for TransAlta. Motivated by this success, the company set a goal of a further 10 percent reduction each year until the goal of zero safety accidents is achieved.

A perfect safety record may be impossible to achieve, but TransAlta people are working toward that goal every day.

Target Zero logo, 2002. TRANSALTA COLLECTION

high frequency) radios in the field by 1953. Calgary Power was a pioneer in adopting both of these systems in Canada.

The second project was a set of three dams at Spray Lakes: Spray, Three Sisters, and Rundle. The dams had a combined capacity of about 65 MW with two-thirds of this at Spray due to its spectacular 275-metre drop. Construction began in 1948 and was completed in 1951. This was Calgary Power's largest hydro project to date and is visible from the Smith-Dorrien Road that runs south from Canmore.

While the Spray Lakes project was under construction in 1949, Calgary Power had to purchase nearly $1 million in additional steam-powered generation due to inadequate water flows—the lowest in forty years. It was a reminder of the risks of relying on nature to supply the water needed for the hydro system. In 1950, a huge snow pack and heavy spring runoff gave the company a bonus when the large Spray reservoir filled a year earlier than expected.

In 1954, Calgary Power installed the third unit, originally scheduled for 1929, at the Ghost plant, bringing the facility up to a total capacity of 51 MW. It also built the small 17 MW Bearspaw Dam on the Bow River, west of the Calgary city limits. The dam helped limit ice jams and winter flooding in Calgary, as well as generate power.

In 1955, Calgary Power commissioned the last dams on the Kananaskis River with the small Pocaterra (15 MW) and the even smaller Interlakes (5 MW) power plants. Each of these hydro plants was minor on its own, but as part of an integrated storage and generation system, it made economic sense. In theory, a cubic metre of water from this area could generate power eight times before reaching Calgary. These developments completed the Bow River watershed system. Max Aitken's vision of tapping the waters of the Bow had been achieved.

Bob Smith in Seebe Control Centre, 1952. TRANSALTA COLLECTION

Interlakes Dam and hydroelectric power plant construction site, Kananaskis, 1955. GLENBOW ARCHIVES NA-4477-18G

Pocaterra Falls, Kananaskis, c. 1930s. TRANSALTA COLLECTION

Pocaterra surge tower, c. 1950s. TRANSALTA COLLECTION

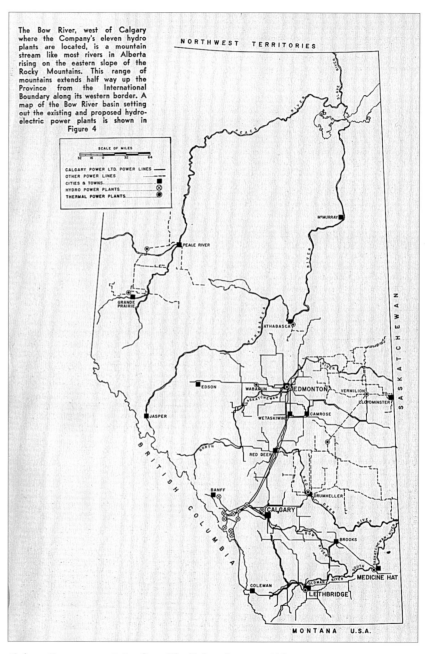

Calgary Power transmission lines. The Relay, *Summer 1957.* TRANSALTA COLLECTION

THE MOVE TO THERMAL POWER   81

Construction of penstock at Spray Lakes, 1948. TRANSALTA COLLECTION

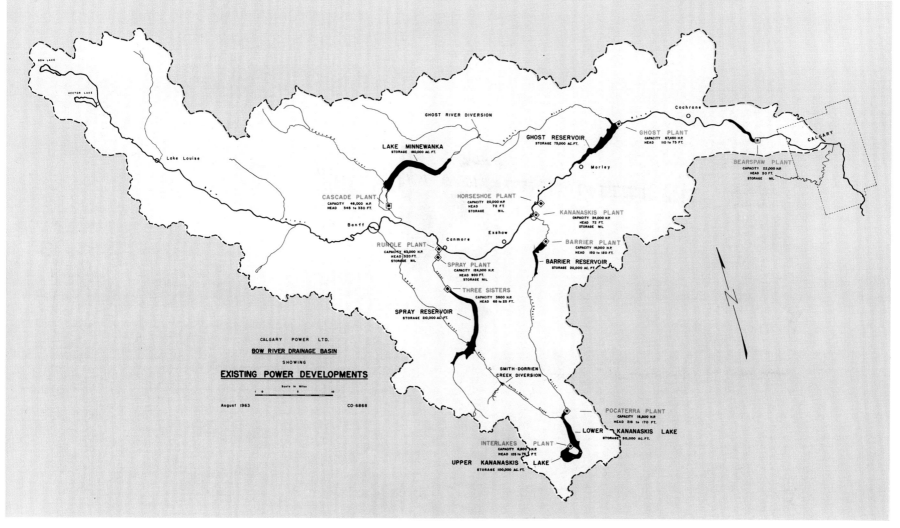

Map of Calgary Power hydro sites, "Power for Progress" booklet, 1968. TRANSALTA COLLECTION

A Seamless Succession

The upsurge in the Alberta economy had allowed Killam and Gaherty to move aggressively on these new hydro projects and seek opportunities for power generation elsewhere in the province. Calgary Power had also begun restructuring its operations in the late 1940s when it moved its head office from Montreal to Calgary. Now the company was functioning under a three-tiered structure, with its corporate headquarters in Calgary; regional offices in Lethbridge, Canmore, Calgary, Red Deer, Edmonton, and Wetaskawin; and twenty-four district offices spread across the province.

Change came quickly after Killam's death in 1955. He had owned the company since 1920, and his connections in the financial world would be difficult to replace. The relationship with Royal Securities—

continued on page 85

Bert Howard at Bearspaw Dam, 1955. TRANSALTA COLLECTION

Bert Howard

Born in Calgary in 1913, Bert Howard grew up as the son of a trust company manager. He attended Earl Grey Public School and then the Central Collegiate Institute. Like many other leaders at Calgary Power, he worked for Montreal Engineering in New Brunswick, Nova Scotia, Prince Edward Island, and Newfoundland before returning to his native Alberta.

He became general manager of Calgary Power in 1952 and vice president in 1959. As president of the company from 1965 to 1974, he oversaw the development of coal-fired generating plants during Alberta's largest boom to that time. In 1973, he became chairman of the board, a position he held until 1984. At his retirement party, the company gave him his old floor-sweeping broom, the one he used at Seebe, as a going-away present. Or at least that's how the story goes.

"The company has been innovative," Howard said of TransAlta on its seventy-fifth anniversary. "It has been forthright in its conduct with the public, it has been a forerunner in power systems technology, and it has been far-sighted in its planning. But, probably most important of all, it has developed an ongoing commitment to the needs of its investors, its employees, its customers, and to the communities it serves."

In his retirement, Howard continued to contribute to his community in many ways, and he served as president of the Canadian Nuclear Association from June 1976 to June 1977. In 1997, the year after his death, he was honoured with a plaque at the University of Toronto's Engineering Alumni Hall of Distinction. The plaque reads: "His significant contributions to Canada's electrical grid system are his legacy."

Employees in front of Calgary Power trucks. L to R: George Garner (district manager, Olds district), Ray Laskoski, Gordon McColl, and Dick Sandilands (patrolmen), 1960.
TRANSALTA COLLECTION

so important to Calgary Power's financial health—weakened, especially after Royal Securities merged with a Canadian and then a U.S. brokerage firm in 1969 and no longer financed all of Calgary Power's operations. And although the Montreal Engineering connection continued into the 1990s, it moved from controlling the technology to being principal advisor and consultant.

While Gaherty continued on as president and became chairman of the board in 1960, a more locally developed leadership, under Harry Thompson and Bert Howard, evolved in the 1960s and beyond. Thompson was president from 1960 to 1965, and when he moved on to become board chairman, Howard replaced him as president. The Thompson-Howard team would direct affairs for two decades. It was a seamless succession that brought new skills and experience to the leadership of the company, particularly relevant during the upcoming addition of thermal generation to the company's hydro operations.

Killam's death brought about another major change for the company. For the first time in its history, its shares were traded on the stock market, and the company became a widely held investment. But before Calgary Power offered shares to the general public, it allowed its own people to take an ownership stake in the company. More than one-third of Calgary Power employees took advantage of the offer. The rest of the shares sold quickly in the public offering at $60 per share. The shareholder base expanded to more than ten thousand individuals holding common or preferred shares. As a result, the Toronto Stock Exchange considered Calgary Power shares a preferred blue chip investment, and the company's fortunes attracted attention in the *Financial Post* and other financial papers.

Cause for Celebration

On 21 May 1961, Calgary Power paused amid its expansion plans to celebrate the fiftieth anniversary of the delivery of the first power into the Calgary grid. It was a time to remember the company's achievements, to honour those responsible for its success, and to give back to the province that had made it all possible.

The company celebrated its Golden Anniversary with three projects. First, it published a heavily illustrated series of articles in *The Relay* in 1961 and 1962, reviewing the major events of the first five decades of Calgary Power's history. Calgary Power's Public Information and Consumer Service division worked with James Lovick & Company to create a series of ten advertisements, called the "Face of Service," which ran in Alberta publications and explained the services Calgary Power was providing. Calgary Power's third anniversary project was to sponsor the creation of the Alberta Junior Citizen of the Year Award. The award honoured the achievements of outstanding young Albertans. Each of the six winners was chosen

continued on page 90

Hiring for the Future: Employees Who Became Leaders

L to R: Marshall Williams, Bert Howard, Harry Schaefer, and Walter Saponja. Bert Howard retirement party, 1984. The plaque is a retirement gift and is the name plate of one of the generating stations at Seebe, where Howard's first job was sweeping the floor at the plant. TRANSALTA COLLECTION

As Calgary Power and its assets grew, the number of employees increased along with it. From 400 in 1950, the numbers jumped to 540 in 1955 and to 759 in 1963. During the boom, when the oil patch offered more lucrative salaries, retaining people became a challenge. However, Calgary Power successfully recruited new technicians from the institutes of technology in Calgary and Edmonton and from the engineering schools at Queen's University in Ontario and the University of Alberta in Edmonton. Many of these new recruits would play critical roles in the company's future.

For example, in 1954, Marshall Williams arrived from Nova Scotia to work for Calgary Power via Montreal, where he worked for Montreal Engineering. Football player Walter Saponja joined Calgary Power directly from Engineering at the University of Alberta and began to tackle the challenges of the Wabamun thermal plant in 1961. Ken McCready arrived in 1963 with a degree in electrical engineering and an interest in the economic side of the business. John Tapics came to Alberta from the tiny mining town of Red Lake, Ontario, with a degree from Queen's University, joining Calgary Power in 1978. These young engineers, and many more like them who were hired during this period in the company's history, developed outstanding skills, and their decades of service are part of TransAlta's valuable corporate memory.

Reddy Kilowatt power ad, 1957.
TRANSALTA COLLECTION

Employees' Associations and the IBEW

As early as the 1930s, Calgary Power employees were involved in company decision-making through a program called Employee Representation. According to Mike Halpen, vice-president of human resources from 1969 to 1993, booklets entitled *The Plan of Employee Representation* explained the company's communication and employee consultation program to Calgary Power workers.

On 1 January 1943, Calgary Power implemented a pioneering pension plan for its employees, which was rolled into the Canada Pension Plan in 1966.

In 1945, the federal government's National Wartime Labour Relations Board (created in 1944) certified the Calgary Power Employees' Association as the formal bargaining agent for company workers, and in 1947 the provincial government passed the provincial Labour Act. According to *The Relay*, in 1947, 204 of the company's three hundred employees (management was not eligible) were members of the employees' association. Employees were elected to represent office staff, field staff (north and south of Red Deer), hydro plants, and the Victoria Park generating station. Monthly union dues were 50 cents for men and 25 cents for women. Cash on hand at the end of 1946 was $1,783.49, or more than $20,000 in 2011 currency.

In 1950, Calgary Power employees and those in the subsidiary company, Farm Electric Services Limited, "joined the ranks of the 26,000 self-helpers in the province of Alberta by forming a Hydro Savings and Credit Union Limited," reported *The Relay*. By 1965, the credit union had 771 members of the 960 Calgary Power staff. Hydro Savings was one of 1,900 such institutions in Canada. Loans and insurance, as well as an annual dividend, were available to credit union members.

Also in 1950, the Calgary Power Employees' Association as well as the International Brotherhood of Electrical Workers (IBEW) were certified as bargaining agents for the employees. The employees' association represented all workers at Calgary Power except for those at the generating facilities, who were represented by the IBEW local 254.

Employees and management at TransAlta have always found a way to work out contracts, without resorting to strikes—though they came close a few times. As Harold Taylor recalled of his days negotiating on behalf of the IBEW unionized employees, TransAlta has been "probably one of the better corporations." He and Walter Saponja, who became president of the company in 1996, sat on opposite sides of the bargaining table and grew to appreciate each other." We never had an issue that was serious enough to strike over," Taylor recalled. Their relationship was good, and "we were fortunate enough to bring in mutual gains bargaining." Under this system, both sides present their legitimate interests and approach the issues through listening and trust, in order to work toward sustainable solutions.

On the eve of the company's centennial, Taylor commended it in 2010 for its success: "It's a local company that's done really well; you gotta cheer them on. I can't say anything wrong about them, and they were part of the growth of Alberta."

The Face of Service

the part of us most people see

The Browns have been visiting us — looking at our hydro plants, our transmission lines, our thermal plants, at many of the phases of our work that together form a complete picture of our service. Yet here at home, in their own kitchen, they are seeing and *using* the most familiar expression of electrical energy. The Browns, like a growing number of Albertans, enjoy the modern, fully automatic, convenient kitchen (and other home facilities) that *only* electricity can provide. This is the part of us most people know — in the home doing the countless jobs that make life easier and better.

We are happy to bring you the 20th century servant, electricity, in the most dependable and economical way possible.

No. 8 in a series: Face of Service.

CALGARY POWER LTD.
Serving the province of ALBERTA

You can't shake hands with a generator

Have you ever *felt* sheer power? You would here, just as our touring family, the Browns, are doing, in front of one of the original generating units at the Horseshoe plant. The Horseshoe station was built in 1911 to serve the City of Calgary, and is the oldest plant on the Company's system. To-day, there are 11 hydro-generating plants on the Bow River System, contributing over 300,000 horsepower of electrical energy to serve your personal wishes and the bounding needs of Alberta industry. This one generator, powered by falling water, produces 5,000 horsepower; more than 25 modern automobile engines put together! Dianne Brown realises that a very small part of this power is probably cooking her supper in the electric oven at home this minute.

No. 3 in a series: "Face of Service"

CALGARY POWER LTD.
Serving the province of ALBERTA

The Face of Service

*to see the "Face of Service"

Put yourself in Mr. Brown's place – haven't you ever wondered about your electric bill? What services it represents? Mr. Brown has wondered, and his daughter, Dianne is curious, too. Mrs. Brown, working in that comfortable all-electric kitchen is probably more interested in the results of her electric cooking than in the source. She's interested in people, though – she likes to know who they are and what they do. So, we at Calgary Power have arranged to take this family on tour. We want to show them our company from beginning to end, how it works, what it does. More important, we want them to meet the people who make it work. Join their tour, on this page, in the months ahead.

No. 1 in a series: "Face of Service"

CALGARY POWER LTD.
Serving the province of ALBERTA

Calgary Power is more than dams, transmission lines, substations, machinery — Calgary Power is *people*. Through our touring family, the Browns, we have tried to show you some of these people and the facilities they use to provide plentiful, dependable electrical service. We could not possibly show them all because in order to give you the service you need, we have a team over 800 strong! Together, these 800 people make up our "Face of Service". Together, they are Calgary Power Ltd.

All of them, including General Manager A. W. Howard (who is showing Dianne Brown a map of our electrical system), are devoted to providing the *best possible* electrical service throughout 75,000 square miles of Alberta. And that means the best possible electrical service for you.

This goal, worked for unceasingly by all 800 members of our family, is our contribution to building a Better Alberta.

No. 10 in a series: Face of Service.

CALGARY POWER LTD.
Serving the province of ALBERTA

from submissions made by the public to Alberta's weekly newspaper editors. Verlin Rau of Beiseker, for example, although suffering from a heart condition, performed heroic acts, including saving his brother from serious injury as well as taking in the harvest when his father fell ill. And thirteen-year-old Linda Bochek of Strathmore took care of her family when her mother was in the hospital, while still attending school, taking music lessons, and taking part in a track meet as well as passing her Red Cross Senior swimming test. Advertisements in publications across Alberta honoured the award winners for their achievements.

For Alberta's Diamond Jubilee in 1965, Calgary Power's Public and Employee Relations department commissioned a sixteen-page booklet called *Alberta: Land of Freedom and Opportunity*. In a comic-book format, in full colour, the host—The Man from Alberta—introduces children to explorers, fur traders, the building of the CPR, the arrival of the police, Louis Riel and the rebellion, as well as the creation of the province in 1905. "Beauty and historical spots are depicted; industry and natural resources, natural gas, mining, oil, electric power, agriculture are briefed with up-to-date figures on production and reserves," reported *The Relay*. "The book was carefully researched to provide accurate and up-to-date information and does not carry any advertising messages."

Sample copies were sent out to schools for review, and the Calgary School Board "placed a substantial order for distribution to elementary grades." *continued on page 93*

Calgary Power fiftieth anniversary logo.
TRANSALTA COLLECTION

Linda Bochek from Strathmore, Alberta, was a recipient of the Alberta Junior Citizen of the Year Award, 1962.
TRANSALTA COLLECTION

Booklet produced by Calgary Power to celebrate Alberta's Diamond Jubilee, 1965. TRANSALTA COLLECTION

Our Alberta Heritage

In the early 1970s, Jacques Hamilton, editor of *The Relay*, Calgary Power's employee magazine, embarked on a new publishing project. This was a series of books entitled *Our Alberta Heritage*.

"The history of Alberta has always fascinated me," wrote Calgary Power president G. H. Thompson in his introduction to the first three books in the series, published in 1971. He liked Hamilton's stories of people, progress, and places, as did the public—indeed, so well that the company released another two in the series, about mountains and new pioneers, in 1975. "These qualities—a stubborn, determined independence—are present in the young as well as the old," Hamilton said in a 1972 issue of *The Relay*. "It's the business of how people meet their challenges of their own fates …"

Churchmen is the title of one chapter in the book about people. In it we meet John McDougal in December 1875. The food had run out, and hunting in every direction proved futile. "I well remember one very cold evening as we went into camp my brother shot a coyote. When we had taken off the pelt it looked like good meat." And so they tried roasting it over an open fire. "In lifting a little of the roast to my mouth I caught its odor and my stomach revolted. I at once concluded to fast awhile longer."

Bud Cotton appears in the *Pioneers* chapter. By 1913, all the buffalo had long since disappeared, but when Cotton showed up for duty at the Wainwright Buffalo Reserve, he got a chance to try his skills counting the mighty beasts. Closing in on one part of the herd, they were suddenly charged by an old bull. "The wise saddle horses had apparently known just what to expect. Both ponies wheeled and got out of there pronto. I lost my hat as my spine hit that saddle cantle with a tooth-rattling jolt."

Central Alberta's Annie Gaetz was elderly but still able to tell stories about her immigration to the province from Nova Scotia in 1903. "I had a very bad cough," she recalled, "and went to see the doctor and found out I had tuberculosis." Told to sit still and hope for the best, she rebelled. "That didn't suit me at all, and I thought that if I was going to die I might as well leave and die out here as die there."

In the *Law Makers and Breakers* chapter, a boozer named Joe Perotte in the Crowsnest Pass decided to deal with prohibition his own way: "I'd been drinking beer since I was old enough to jump over a miner's lamp and I'm damned if I'm gonna stop now."

These and many more characters filled the five volumes of *Our Alberta Heritage*, gathered together for the enjoyment and edification of the people of Alberta. Almost fifty thousand copies were distributed through 1975.

Recognizing the popularity of the series, Calgary Power commissioned conservationist and storyteller Andy Russell to tell more stories set in Alberta's past, on the radio. For decades his "Our Alberta Heritage" radio moments recounted the story of the province to Alberta residents. This project had a precedent, as the company first took to the radio waves in the 1950s, with audio versions of its annual report, tales of early Alberta pioneers, and a miscellany on everything from industrial developments to biographies of artists.

G. H. Thompson presenting Our Alberta Heritage *to Alberta's Lieutenant-Governor, Grant MacEwan, 1961.* TRANSALTA COLLECTION

G. H. Thompson

G. H. Thompson was a man who made things happen. Born in Nova Scotia, he earned an engineering degree from McGill University in 1913 and served with the Royal Canadian Engineers during the Great War. He worked for Canadian Westinghouse Company Ltd. before moving west in 1922 to the West Canadian Collieries coal mines. Thompson hired on as assistant superintendent for Calgary Power at Seebe in 1925 and became general manager of the company in 1931, overseeing the construction of hydro developments along the Bow River corridor for several decades. He became president of Calgary Power in 1960, a position he held until 1965 when he turned over the reins of the company to A. W. Howard and moved on to become chairman of the board.

"Harry Thompson spent most of his time chasing literally all over Alberta in a Model T Ford Coupe on the gravel roads," G. H. Milligan recalled in the early 1980s, "signing up all the little towns around—Okotoks, High River, and way down to Lethbridge. Towns up north between Calgary and Edmonton and out east at Camrose. He got franchises with them, and then we started building lines out to those places."

But power lines needed consumers at the end of the wire.

"It was G. H. Thompson's idea," Homer Lebourveau recalled, "trying to get the Home Economics departments in some of the larger schools to put in an electric range."

The company also conducted field trips, where it took Home Economics teachers out to the Ghost Dam. Lebourveau recalled that the company "wanted me to show them around the plant and to explain to them what a transformer is. They had to know that. There were about six of them—teachers from Lacombe, Brooks, and towns of that size—to try to get them interested in the use of electrical appliances. We were putting on shows in schoolrooms or auditoriums."

The construction of the Wabamun thermal plant, one of the lowest cost power producers at the time, signalled a change for Calgary Power and electrical power generation in Alberta. Thompson oversaw the planning, design, and construction of thermal generating facilities in Alberta and Newfoundland, as well as in Central and South America.

In 1964, G. H. Thompson received the Julian C. Smith medal from the Engineering Institute of Canada, "for achievement in the development of Canada."

Thompson died in 1975.

G. H. Thompson. TRANSALTA COLLECTION

King Coal

By the 1950s, the potential for hydro expansion in Alberta had been nearly exhausted. Productive sites in the Bow River watershed had all been developed. Sites on northern rivers, such as the Peace and the Athabasca, were considered and deemed not economically viable when compared to alternatives. Although oil and natural gas were the exciting and much publicized new fuels during the booming 1950s, the company's long-term planners went in a different direction and chose Alberta's other black gold instead. It was an incredibly bold decision: no one had ever mined coal for power on such a large scale.

Most of Canada's supply of coal lies beneath Alberta's wheat fields. Given that natural gas had replaced coal as the fuel of choice in homes and industry, and fuel oil and diesel had replaced coal in railway engines, it seemed that "King Coal" was in decline. The choice to base the future of the electrical utility on coal may have seemed radical at the time, but Gaherty and his colleagues correctly predicted that coal would be less expensive than all other fuel sources for generations to come. They invested heavily in the infrastructure to move into thermal generation.

Securing a site was the first step in developing the coal program. After careful consideration, Calgary Power chose the Lake Wabamun area west of Edmonton. Large seams of sub-bituminous coal rise to the surface in this region, making it ideal for open-pit mining. Although this quality of coal does not produce as much heat as the anthracite coal of the mountains, the area had other advantages: decades of fuel could be mined cheaply, Lake Wabamun water could be used for cooling, and the Canadian National Railway tracks ran nearby, allowing for transportation of materials during construction of the plant. It was only 64 kilometres west of the growing Edmonton market and would have access to the city's large labour pool. As well, the site was close to the major Yellowhead Highway to British Columbia. Exploratory drilling quickly established the thickness of the coal seams and promised a reliable supply for fifty years—in excess of 45 million tonnes. Forest and marginal farmland covered the area that would be mined. There were a few cottages on the lake, and a First Nations reserve some distance to the east.

The Mannix family owned the coal reserves and asked a high price for the rights. Marshall Williams, then a junior executive with Calgary Power, negotiated the coal contract with Mannix in 1955, purchasing the shares of Alberta Southern Coal Company Ltd. To sweeten the deal, Calgary Power offered Mannix the contract to mine the coal. After lengthy negotiations, Calgary Power agreed to pay about one cent per tonne for the coal in the ground and negotiated a low royalty rate with the province. The company's strategy relied on low fuel costs, and this advantage became the centrepiece for its thermal power generation for the next half century.

Calgary Power entered the coal market at an ideal time. Although the Alberta government favoured the export of petroleum to continental markets, it discouraged the sale of natural gas outside of provincial boundaries. The highest and best use for gas, according to the leaders of the day, was as a home heating and cooking fuel, which in a cold country was an emotional argument that carried the day.

Coal, as a result, was a perfect fuel—plentiful, cheap, and not controversial. As the price fell, utilities across North America adopted this resource. The huge thermal coal reserves apparently had few other uses. The exodus of workers from coal-mining towns in Alberta also meant new jobs at Wabamun for the unemployed miners.

Gaherty watched these developments carefully, and, given his love of hydro, took his time to be won over to the benefits of thermal generation. As an experienced executive at Montreal Engineering, he knew well the technical issues and the risks involved with coal. But, as Calgary Power ran out of locations to develop new hydro plants, Gaherty was forced to face the inevitable: the company would have to broaden its fuel base in order to meet rising customer demand.

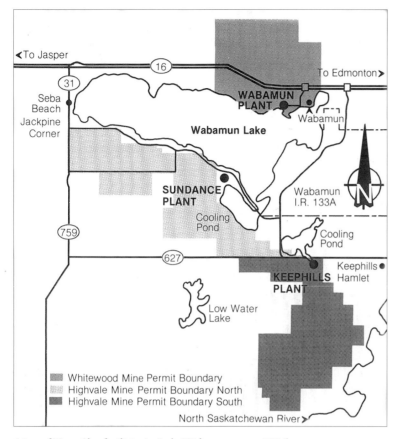

Map of TransAlta facilities in Lake Wabamun area, 1986. TRANSALTA COLLECTION

Road construction during building of the Wabamun plant, early 1950s.
TRANSALTA COLLECTION

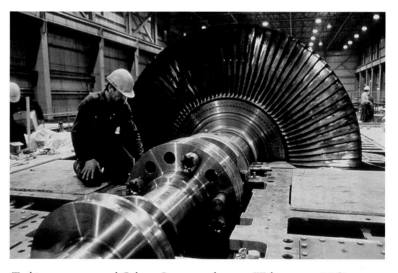

Turbine generator and Calgary Power employees at Wabamun, c. 1960s.
TRANSALTA COLLECTION

The first Wabamun unit was commissioned in 1956 at a cost of $8 million. The second, two years later, cost $7 million. Both had a 66 MW capacity, compared to the standard 400 MW units of today. But in their day they represented the latest in thermal technology and were Alberta's largest thermal power plants. The company organized tours to show off the new equipment to an interested public. The generating capacity of these first units was about 60 percent of the existing hydro units that they augmented. Because cheap natural gas was available on short-term contracts, Calgary Power initially operated

Opening at Brazeau, 8 October 1969. Marshall Williams is pictured (circled) standing by the dam. TRANSALTA COLLECTION

these thermal units on gas until its mining operations were capable of supplying plentiful and less costly coal. The company also built a major transmission line from Wabamun into the Edmonton market.

Alberta power consumption was growing more than 10 percent per year—doubling every seven years. The company added two much larger units in 1962 and 1968. Unit #3 was 147 MWs—nearly double the original installations—and unit #4 was nearly double that, at 286 MWs. Calgary Power was able to keep costs low with improved economies of scale and plant efficiencies.

Coal technology was complex and required more maintenance and surveillance than hydro to ensure reliable operations. Hydro was a simpler operation, but it required an adequate flow of water. Using both hydro and coal to generate power was both pragmatic and innovative. Reliability is a foundational principle in the utility business, and the synergy of hydro and coal allowed Calgary Power to supply clients and meet economic targets.

Calgary Power was forced to recruit personnel with a new skill set when it expanded its operations into coal. Some operators required a technical school diploma in power engineering, while others needed a chemical diploma. There was also a need for specialized professional engineers. Walter Saponja, who joined the company in 1961 and later became its president, loved the meticulous work of continuously improving the technology—gaining one cent of productivity with one adjustment, two cents with another. The Wabamun plant's efficiency was outstanding, and the company won awards for the highest labour efficiency in North America.

continued on page 98

Mining the Coal

Calgary Power's involvement with coal mining started in 1962 when it began using coal to generate electricity at Unit 3 at Wabamun. The first supply of coal came from the Whitewood Mine, just north of Lake Wabamun. The capital-intensive project supplied 2.8 million tonnes of coal annually and utilized a walking dragline with a 106-metre boom and 60-cubic-metre bucket called the "Spirit of Whitewood."

The company also owned the electric shovels, front-end loaders, coal hunters, scrapers, bulldozers, and ash haulers. Its strategy in owning the equipment was to control the ability to mine in case of a dispute with the mining contractor.

The mining process involves a number of steps, each with its own routines. First, huge earthmoving equipment remove the topsoil and subsoil with their valuable nutrients (this soil is put aside for reclamation). Then the giant walking dragline removes the overburden to expose the coal seam. The overburden ranges from 6 metres to 37 metres, and there are usually five or six seams of mineable coal—a crumbly sub-bituminous coal—ranging from 30 centimetres to 3.5 metres in thickness. At this stage, large electric shovels and front-end loaders pile the coal into bottom-dump coal haulers, which convey it the short distance to the power plant.

After a coal seam is removed, shovels and bulldozers clear the interburden between seams. When all the accessible seams are exhausted, the overburden is returned to fill the pit.

As a final stage, topsoil is returned so that vegetation can re-grow. Three years later, the reclaimed areas are ready for the provincial inspectors.

The coal is cut row by row, so reclamation begins after each row is completed to minimize the visual impact of the operation. As of 2011, TransAlta has reclaimed more than 75 percent, or 1,070 hectares, of the 1,822 hectares of land mined at Whitewood. Reclamation began in the early 1960s, before the Alberta government instituted the Coal Mines Regulation Act and the Surface Reclamation Act in 1963, and the Land Surface Conservation and Reclamation Act in 1973.

> Environment wasn't even a word hardly when we began reclaiming the mined-out land. There weren't half the number of regulations there are now, but we said, "We feel as good citizens of the province, we should not leave ugly scars around the place. We should reclaim, we should make a good effort to put it back in the same state or better when we are finished."
>
> –Tim Finnis, Calgary Power employee

Coal enters the plant on a conveyer belt, where a series of pulverizers crush it into a flour-like mixture. Hot air then blows the fine coal particles into the combustion chamber of the boiler where it ignites at temperatures reaching 1300°C.

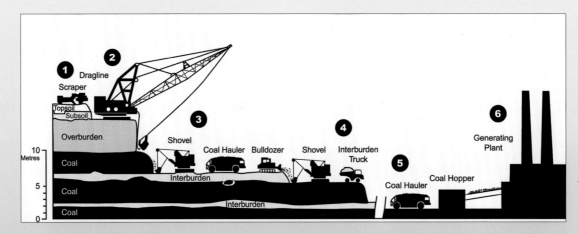

The mining process, 2004.
TRANSALTA COLLECTION

The boiler is lined with water-filled tubes, and the heat converts the liquid into steam. Smoke from the coal combustions passes through precipitators that remove over 99 percent of the particles—called fly ash—as it exits the stack. High-pressure steam enters the turbine, forcing circular plates with slated holes to rotate the shaft. This shaft then drives the generators to create the electricity. A water purification plant provides chemically pure water for the boilers so there is no scaling in the turbines. Then a condenser returns the steam to a liquid form. Water from Lake Wabamun absorbs the heat from the condensers. The ash-handling systems transport the ash waste, some to a slurring lagoon and some back to the mined-out area. Some of the fly ash is sold to cement companies and used as filler in making concrete. The electricity from the generators goes to a substation where a transformer increases the voltage in order to allow for long-range transmission to distant markets.

Calgary Power's first dragline, a Ransomes Rapier 1350, 1960. TRANSALTA COLLECTION

The Bighorn Dam, c. 1970s. TRANSALTA COLLECTION

Cooperating with the Province

In the early 1960s, Calgary Power explored hydro sites on the North Saskatchewan River, west of Edmonton. The sites were not financially feasible, so the company continued to expand development of thermal power at Lake Wabamun. However, the Alberta Power Commission asked Calgary Power to consider a hydro site on a tributary to the North Saskatchewan, the Brazeau River. This would help the province control seasonal water flows in Edmonton that could result in "a serious pollution problem" with sewage. In exchange for collaborating on the project, the province funded the construction of the dam and water storage area—which became operational at the Big Bend site on the Brazeau River in 1963—while Calgary Power paid for and installed the generating equipment, which began providing electricity into Calgary Power's transmission system in 1965. When fully operational it generated 355 MWs—more than all the Bow River watershed plants combined. However, the limited flow of the Brazeau and the limited storage capacity of the dam meant that it could only provide peaking power for short periods of high demand. In this way, it complemented the base load coal-fired plant at Lake Wabamun.

The company and the government worked together on a similar project at the Bighorn site on the North Saskatchewan River and commissioned it in 1972. It had less head, more flow, and much greater potential for storage. It generated 120 MWs. Though the capacity was one third of Brazeau, the Bighorn Dam created the largest reservoir in the Calgary Power system, the thirty-kilometre-long Abraham Lake. Given these water resources, the Bighorn generating plant was able to operate as base load for the regional customers.

According to a 2010 report by the Alberta Utilities Commission, these unique projects have served the City of Edmonton's water supply and sewage disposal needs, provided cooling water for the city's thermal plants, and supplied water for the refineries and chemical plants in the industrial area east of Edmonton.

Sundance

In the late 1960s, as Calgary Power installed the last unit at Wabamun, plans were in place to build new capacity for the 1970s. Engineers began designing the largest project in the company's history, the Sundance complex on the western shore of Lake Wabamun. It would utilize six generating units totalling over 2000 MW, using pulverized coal technology. Coal would come from the Highvale Mine, south of Lake Wabamun, and cooling water from the lake or from the North Saskatchewan River.

The first unit opened in 1970, and the others followed at regular intervals until the company commissioned the sixth unit in 1980. The plant soared nearly thirty storeys above the shoreline of the lake and, with the latest technology, was the pride of the fleet. The complex was the second-largest thermal power plant in Canada, second only to Ontario Hydro's Nanticoke on Lake Erie, which was the larg-

Construction at Sundance, 1976. TRANSALTA COLLECTION

est in North America. The Sundance unit also boasted some of the best cost and efficiency ratings on the continent. Officials from all over North America came to see it operate.

Low-cost coal and the efficiency of the Highvale Mine were key factors in Sundance's success. Its coal production became the largest in Canada—about 11 million tonnes per year, and its operating and coal reserve areas covered 120 square kilometres, with the initial estimate of 560 million tonnes of recoverable coal. Alberta's coal is extremely low in sulphur; this advantage proved critical when governments began setting regulations to reverse the effects of the acid rain caused by sulphur dioxide.

A Change in Government

After thirty-six years in power, the Social Credit party lost control of the provincial legislature in 1971 when Peter Lougheed and the Progressive Conservative party took office. The SoCreds had experienced the hard times of the 1930s, followed by the booms caused by oil discoveries, and had formulated an industrial development policy where the firm hand of a powerful government guided the activities of private companies as they employed investor funding to exploit the province's bountiful natural resources. In fact, Premier Ernest Manning believed cheap nuclear energy would eventually render Alberta's petroleum birthright worthless, and encouraged the efficient production and export of oil and gas as quickly as possible by Canadian and multinational corporations.

The next premier was the grandson of Sir James Lougheed, a great proponent of western rights who became a Canadian senator. The Lougheed family had suffered during the Depression, losing its grand sandstone family home in Calgary. Peter Lougheed and his colleagues in the Progressive Conservative government that took power in 1971 respected the SoCred accomplishments, but they wanted more. This government's Fort McMurray Tar Sands Strategy, published in 1972, pinpointed the ideological shift:

> On one hand we can continue the policies of the conventional crude oil developments creating tremendous and unregulated growth and developments resulting in short term benefits … in addition to the depletion of non-renewable resources. Conversely we can regulate the orderly growth and development of the bituminous tar sands for the ultimate benefit of Alberta and Canada in order that Canadian technology will be expanded, Albertans will find beneficial and satisfying employment within its diversified economy, and our environment will be protected and enhanced for future use.

The definition of the "public good" was changing, and though the Lougheed government was not radical enough for some during the turbulent days of the 1970s, the oil patch nicknamed the premier "the blue-eyed sheik" for the powerful way that he directed the industrial development process in Alberta from 1971 to 1985. Where Manning had worried that petroleum would lose its value, Lougheed feared the resource would be exploited so quickly that it would run out. As a result, he was not unwilling to confront power—in the form of federal politicians, the presidents of oil companies, or anyone else who stood in his way. The authors of the 1972 report warned that: "Conflict will arise when the principles of government and the individual corporation do not coincide."

Coincidentally, Calgary Power's operations became more complicated. Provincial regulatory boards, formed by previous governments, began requiring public hearings, expert testimony, and public consultation. Although the company always filed its plans with the appropriate agencies, and those plans were accepted, sometimes citizen groups challenged projects, the hearings became lengthy, the number of experts called to testify grew long, and the paperwork grew exponentially.

In the early 1970s, Calgary Power also had to request rate increases for the first time—an unprecedented development given that, for the first sixty years of its history, the price it charged for electricity had only ever decreased. Environmental assessments were another requirement of the 1970s, adding more work for employees as they prepared material and reports for these hearings and defended the company's development plans. This was also a period of high inflation in Canada. Spiralling wages, higher interest rates, and increasing construction costs and materials hurt the company and its staff.

Load and Sales

Calgary Power's load in 1974 was 8,631 million kilowatt hours or about 55 per cent of the total interconnected grid load in the province. Sales totalled 7,952 million kilowatt hours.

Load in Kilowatt Hours

	1969 millions	1974 millions
Total System Load	8,093	13,567
Calgary Power Ltd.	5,219	8,631
% CP Ltd. of Total Province		55
% Increase CP Ltd. Load 1969-1974		65

TRANSALTA COLLECTION

The Shock of Camrose-Ryley

The coal plants at Lake Wabamun raised distinct environmental issues for Calgary Power. Some cottagers in the area objected to the industrial development that altered the landscape around the lake. As environmental consciousness increased during the 1960s, air and water quality issues became more important, and citizen involvement, including public protests, became common. The company also faced rising scrutiny from consumer associations that challenged the prices at which power was sold to the public. Calgary Power and its president, Marshall Williams, were heading into a new age of environmental consciousness and public transparency.

In the mid 1970s, as the Sundance generating units were being installed at Lake Wabamun, Calgary Power began looking for another source of coal. Drilling and surface exploration pointed to a site in the Camrose-Ryley area—also called Dodds-Round Hill—southeast of Edmonton.

In 1974, Calgary Power applied to the Energy Resources Conservation Board to build a power plant at Camrose-Ryley. Much as they had twenty years earlier, company engineers found a good source of coal, close enough to the surface to mine for forty-five years, and set about planning a development that would eventually involve open-pit mining over more than 100 square kilometres of land.

The huge $2.6 billion project, in partnership with CPR Minerals, would include a 2250 MW power plant and a petrochemical complex, as well as a 96-kilometre water pipeline from the North Saskatchewan River. A booming economy, nearby rail lines, local jobs and tax benefits, and customers for the electricity seemed to assure approval of the project and acceptance by the community.

But much had changed since the company developed its first massive coal mines and thermal generation plants at Wabamun Lake in the 1950s. Innocently enough, Calgary Power promised to reclaim the land after the mining process. Having never conducted extensive public consultations, it saw no need for them now. It could not have been more wrong. The Camrose-Ryley area had been farmed for generations. Prime agricultural land, it was precious to the locals, who questioned the company's ability to restore the land to agricultural use after the mining work was completed.

The Progressive Conservative government released a coal development policy in 1974 that required public disclosure of development plans in the early stages of planning—compared with the former process that revealed plans near the end of the approval process. The government had also raised royalty rates paid on coal—as it did on petroleum, too. It wanted to diversify the economy to ensure continued prosperity when petroleum development inevitably peaked; the mine, power facility, and associated petrochemical plant could have been the first such project.

Local citizens formed the Round Hill–Dodds Agricultural Protective Association and their protests attracted media attention. With 140 members, the association hired experts, one of whom claimed the company's reclamation methods were "like putting makeup on a corpse." The uproar also garnered the attention of the member of the legislative assembly for the region, the premier, and many others. In August 1976, the government announced its decision to refuse approval for the project. Calgary Power immediately began the search for coal deposits elsewhere.

This setback at Camrose-Ryley was a first for Calgary Power. As a company that had always prided itself on good community relations, it had been so preoccupied building generating capacity and expanding its service to customers during the booming 1970s that it had failed to keep up with the changing social and political climate. It was a lesson the company would take to heart. Calgary Power took a new stance, and would come to use a very different public consultation and development process later in the decade when negotiating over the hamlet of Keephills, near Wabamun.

The three decades after the discovery of oil at Leduc represented a "golden age" for Calgary Power, during which it built the company's future. The boom in the Alberta economy had carried Calgary Power along on a wave of growth, with generation capacity increasing approximately twenty-nine times, or almost tripling every decade from 1950 to 1980. Profits by the end of this period rose in line with the company's new assets, nearly twenty-seven times that of 1950, adjusted for inflation.

Calgary Power remained successful by responding quickly and innovatively to change: developing fuel alternatives, including hydro capacity in collaborative projects with the Alberta government, and moving into thermal generation capacity, timing the new plants nearly perfectly to meet market demand.

The company also began to move away from its faithful backer, Royal Securities, as the focus of finance shifted from Montreal and London to Toronto and New York. Calgary Power became an investor-owned utility, with new partners and money markets.

As the Killam-Gaherty era passed into history, management became firmly rooted in the west. In these decades, as in earlier ones, Calgary Power grew along with the province, evolving as a company as the province moved from a primarily agricultural economy to a new focus on urban, oil and gas, and industrial sectors. Within the context of this rapidly changing political, social, and regulatory environment, Calgary Power began to focus on a new sustainable approach to environmental concerns. Above all, the company took pride in contributing greatly to the economic development of Alberta.

Calgary Power logo, 1970s. TRANSALTA COLLECTION

> "Success calls for the removal of fear and this, in turn, requires tolerance for learning by management." MARSHALL WILLIAMS, 1988

CHAPTER SIX

Recession and Market Turbulence

Alberta's economic history is a story of booms and busts. These wild cycles have wreaked havoc on corporate planners and their investment decisions. An electrical generating facility planned during a boom period could be commissioned six years later, when the economy went into recession and the market was over-supplied with power. In the early 1980s, after three decades of rapid growth, Calgary Power—like the province—was unprepared for a sharp decline. Converting from a mindset of rapid growth to retrenchment is a psychological challenge; it was like 1929 all over again.

The decade of the 1980s ushered in a period of exceptional change for Calgary Power and for Alberta's electrical utility industry as a whole. The company experienced poor market conditions and continuing uncertainty in public policy. And new fuel sources emerged to challenge King Coal.

Calgary Power's operational strategy changed from rapid growth to cost containment and customer retention. The years 1980 to 1995 saw the beginning of new commitments for the company beyond the borders of Alberta and the limits of a provincially regulated marketplace.

Marshall Williams, as president, continued to provide steady leadership, building on Calgary Power's legacy: financial strength, quality people, and excellent technology. His work to keep costs down paid off in competitive dividends. Fortunately, the company still had an extensive transmission system and the lowest cost base of

TransAlta head office building, c. 1980s. TRANSALTA COLLECTION any electrical generation system in Alberta. Even in a depressed economy and poor market conditions, the company managed by 1985 to expand its market share to 81 percent of provincial needs.

In addition to the major economic shocks, political developments of the 1980s also affected the Alberta economy. The rapid rise in world oil prices during the 1970s had put pressure on Ottawa to collect more economic rent from the profitable oil industry in order to support federal programs across the country. As a result, the federal government announced the National Energy Program (NEP) on 28 October 1980 to insulate Canadians from high world oil prices with a lower Canadian oil price. This was to be achieved by subsidizing Canadian companies working in the frontier regions and diverting more tax revenue to federal programs from provincial coffers and the oil company bank accounts. The only benefit to the electricity sector was a policy to divert the home heating market from fuel oil to electricity as a means of curbing emissions.

The NEP assumed the price of world oil would continue to climb. It had risen from $2.45 in 1970 to a peak of $44.66 in 1980; some analysts predicted it would top $100 a barrel. But when prices collapsed to $19.10 in 1988 so did the economic rents that they were designed to tap. As economic activity slowed and the rate of inflation rose, the 10 percent annual growth in demand for electricity disappeared, and power companies scrambled to find customers. The economy did not begin to turn around until the 1990s.

The NEP built on historic regional tensions—East versus West; provincial rights versus federal power—and had wide policy implications beyond the energy sector. It created a bitter competitive rivalry, one with serious repercussions for Calgary Power.

Probably at no time in its history has Canada faced greater political and economic turmoil—much of it brought about by federal/provincial confrontations on energy pricing and constitutional reform. Federal budgetary measures and energy policies announced in October 1980 penalize Alberta's oil and gas sector and have an indirect impact on our industry. The resulting indecision and delay on major energy projects undoubtedly will affect future energy sales and planning unfavorably.
– Calgary Power 1980 Annual Report

With the Alberta electricity market under serious strain, a longer-term view became central to the success of Calgary Power. To that end, it made a strategic decision to branch out of Alberta to new Canadian, continental, and international opportunities where it could increase its market share and return on investment. Internally, the company set out to define what it could do best—generation, transmission, or distribution—and it then moved its business focus in that direction. It evolved from an integrated utility to a sector-specific model. This strategy fundamentally changed the structure of the company and required that management anticipate change. "Success calls for the removal of fear and this, in turn, requires tolerance for learning by management," Marshall Williams remarked in 1988, three years after passing on the presidency of the company to Ken McCready.

What's In a Name?

As the shift in strategy was taking place, Calgary Power also examined its moniker. A corporate name is central to a corporate identity. A company's name needs to reflect the nature of its business and its corporate vision and values. It can also reflect the geographical extent of the operations. Above all, it should help drive public recognition and respect for the company in the market and with the public.

In the beginning, the Calgary Power name represented the commitment of a company with its head office in Montreal to serve the people of Calgary and southern Alberta. As a brand, it worked well for decades, and many employees felt a warm loyalty to the grand old "Calgary Power" insignia. But by 1980, half a million customers in Alberta—people and companies—flicked on lights each day from the company's expanding system. Its operations encompassed 194,000 square kilometres from the border with Montana at the south to 185 kilometres north of Edmonton.

In 1981, after several years of discussion, Calgary Power became TransAlta Utilities Corporation. According to board chairman, A. W. Howard, "The change reflects the company's province-wide role in the utility sector and its expanding interest in non-utility operations." Marshall Williams pointed to the company's plans to invest elsewhere in the province and outside of Alberta. A vigorous advertising and education campaign accompanied the name change and helped create public recognition and ensure customer loyalty. "The new name does not alter in any way the Corporation's long-standing commitment of the past 70 years," assured the 1981 Annual Report, "to supply reliable and economic electric power service to Albertans."

During the early 1980s, TransAlta expanded its headquarters buildings at its Eleventh Avenue site in Calgary. It had purchased the entire city block at Centre Street and Eleventh Avenue Southwest during the early sixties, when land prices in that part of Calgary were low. Located south of the main financial section, and across the railway tracks, it was not considered a desirable area at the time, but would prove to be a wise investment as the district developed commercially. The company had built a ten-storey office building on the site in the late 1960s. In 1980, it began construction of a second building, which was reduced, as a result of the economic downturn, from eleven to six storeys by the time the company moved in during 1984.

Employees change the company signs from Calgary Power to TransAlta Utilities, 1981. TRANSALTA COLLECTION

Aerial view of second TransAlta building, built in 1980. TRANSALTA COLLECTION

continued on page 107

Marshall Williams

Marshall Williams. TRANSALTA COLLECTION

"It goes way back to old man Killam," Marshall Williams recalled in 2009, "that your main purpose is to serve the customers … If you serve them well, the shareholders will do well."

Born in Londonderry, Nova Scotia, Marshall grew up in West River Station, Nova Scotia, where his father was the Canadian National Railway agent. He attended a one-room school until Grade 10, "skiing there in the winter and hiking in the summer," and then attended high school in Truro, going in on the train in the morning and back at night, arriving home around 9:00 or 9:30 P.M.

Marshall attended Acadia University and then went to Nova Scotia Tech for a master's degree in engineering before heading off, as had so many from the Maritimes, to seek his fortune elsewhere. The year was 1948.

"Typical Nova Scotian, I owed so much money that I had to hitchhike to Montreal to get a job. When I got there I knew that Montreal Engineering was pretty well all Nova Scotians."

Izaak Killam was the legendary Maritimer who owned Montreal Engineering, as well as Royal Securities, Newfoundland Light & Power, Maritime Electric, Nova Scotia Light & Power, Mersey Paper, Ottawa Valley Power, Calgary Power, B.C. Pulp, and companies throughout South America.

"I knocked on their door and met Mr. Stairs and Dr. Gaherty, and they hired me on the spot." Marshall worked for Montreal Engineering for the next six years, on projects in Newfoundland and Quebec, and he eventually did research into the opportunities for additional hydro development in the west.

In 1954, Marshall transferred to Calgary to begin management training with Calgary Power. He served as assistant to the general manager, Bert Howard. He studied the hydrology of the Bow River and the North Saskatchewan River, and, as a result of its drive to find more hydro opportunities the company eventually built the Brazeau and Bighorn power plants on the more northerly river.

At Wabamun, the company's first thermal power plant, Marshall negotiated the purchase of the coal rights from the Canadian Pacific and assumed the provincial coal leases, a deal that provided a long-term supply of cheap fuel. But the first unit operated on inexpensive natural gas, purchased from the Midwest Industrial Gas wells near Edmonton. Keeping its options open was important to Calgary Power, and strategic planning was a priority. "As the volume of fuel builds up, the cost of coal will go down, and somewhere there was going to be a crossover. So that plant was built so that it could use both coal and natural gas, engineering-wise."

Marshall was appointed a director of Calgary Power in 1972, and president and chief executive officer in 1980. In 1984, he became chairman of the board, president, and CEO of TransAlta Corporation, In 1991, Marshall retired from his office as chairman of the board. He has also served society as chairman of the Banff Centre and was the founding president of the Society, Environment & Energy Development Studies (SEEDS) Foundation.

Hostile Suitors

The newly named TransAlta fought for its very existence when ATCO Corporation attempted a hostile takeover of the utility. This dramatic chapter in the company's story revolved around two very able and determined corporate generals: Ron Southern of ATCO and Marshall Williams of TransAlta.

The battle began with some mild skirmishing. In 1980, when the U.S. corporation that owned 58 percent of Canadian Utilities Company offered to sell its shares, Calgary Power made a bid for them and for the remaining shares held by the public. At that point, Calgary-based ATCO jumped in and purchased the larger block of shares, but Calgary Power successfully acquired the public shares, or 32 percent. A couple of months later, when ATCO launched a bid for all of Calgary Power shares, the staid old electricity sector began generating lengthy columns in the financial press with a very public battle of charges and counter-charges. However, the Alberta Public Utilities Board (PUB) blocked the bid until a hearing could be held.

When ATCO claimed to be beyond the PUB's control, TransAlta—as it was now called—challenged its rival all the way to the Supreme Court of Canada. The court confirmed the Alberta government regulator's authority and, in the end, the PUB prevented ATCO from buying out TransAlta without regulatory approval.

Another suitor made a play for a minority interest in TransAlta in 1981 when real estate developer Nu-West Development Corporation sought and obtained regulatory approval to acquire a 21 percent interest in TransAlta. Though it successfully purchased 21 percent of the company's shares, Nu-West fell on hard times during the economic downturn of the 1980s and was forced to divest its shares of TransAlta in 1982.

With the introduction of Nu-West's purchase of shares, ATCO could no longer bid in an uncontested manner. When Nu-West agreed to sell its shares in 1982, and, with ATCO the only logical

Sheerness power plant near Hanna, Alberta, 1985. TRANSALTA COLLECTION

continued on page 110

Keephills community sign.
TRANSALTA COLLECTION

Moving a Community

Moving a whole community might seem like a massive undertaking, but that is what TransAlta did in the early 1980s. The quiet hamlet of Keephills, 80 kilometres west of Edmonton, stood on the land slated for a proposed expansion of the Highvale coal mine, the source of fuel for the massive thermal generating plants at Lake Wabamun.

Public consultation began in January 1977, and company officials listened carefully to community concerns. The Committee on the Keephills Environment—C.O.K.E.—represented the locals. About five hundred people resided in the affected area, on ranches, farms, and in the eighty-year-old hamlet. Consisting of a school and teacherage, community hall, and four houses, Keephills was not large, but it was an important regional centre and, of course, home to those who lived there.

Talks continued into 1981. Finally, in a spirit of cooperation and collaboration, the company incurred the cost of $4 million to move and rebuild the hamlet. In 1982, the new Keephills opened in a new location on a creek. It included thirty-one residential lots, a new community hall, and a twenty-eight-acre environmental reserve. The four residents of the old hamlet received lots and fully serviced homes in the new community. School enrolment jumped from thirty-seven at the old school to eighty-three by the fall of 1988.

As the Prokop study of the Keephills project—undertaken by the Canadian Environmental Assessment Research Council in 1987—noted, listening takes time. Both sides worked hard to articulate principles and guidelines for the relocation process. "TransAlta drafted a position paper on this subject matter which expressed the company's perspective. C.O.K.E. in turn drafted the community's position paper, which had a very different view. This process was repeated five times as each participant did not respond directly to the other's position paper

Birds-eye view of Keephills hamlet, 1981. TRANSALTA COLLECTION

but instead to their specific interpretation of it. The fifth and final draft was acceptable to both TransAlta and C.O.K.E."

Change is always discomfiting, but the Keephills hamlet relocation case study proves that government, industry, and citizens can work together to create a positive outcome. "Residents we interviewed stated satisfaction with the involvement process ('as good as can be expected given the reality of the situation') and would recommend such a process for other communities, with modification," concluded the Prokop Report.

But living near a major utility project can be challenging, and twenty years later the Keephills project was once again in the news. The 2007 Alberta Energy and Utilities Board (EUB) inquiry into complaints about the coal-mining project found that "TransAlta has lost the trust and goodwill of the community." Notably, Keephills residents also criticized the EUB and Alberta Environment for lack of proper consultation with the local community, for their failure to deal with complaints, and for the lack of enforcement of government regulations on issues including "noise, dust, road closures, water, land purchase or leasing, and trespass." Obviously, there was room for improvement.

TransAlta vice president of Western Canada Operations, Will Bridge, took responsibility for the company's actions and for the fact that the company had "missed opportunities to establish and maintain the trust of the community." Bridge identified structural issues within the company that contributed to the communications problems and "proposed steps to begin repairing the trust issue with the community." These included more meetings with locals, a review of the terms of reference for the multi-stakeholder steering committee, a telephone hotline for ongoing operations issues, and two new hotlines for issues associated with construction of the new mine.

buyer, TransAlta and ATCO came to an agreement to each divest of their holdings in each other's companies. This sensible solution finally eased tensions between the companies. They cooperated in divesting their shareholdings in each other and later partnered in the construction and operation of the Sheerness Power Plant near Hanna, Alberta.

These hostile takeover attempts had brought out the tough business instincts of TransAlta's executives and their skills in defending the firm's market share—skills that would be called on again in the future.

The Keephills Project

As TransAlta looked for further opportunities for growth, it decided to pursue new generation opportunities in the Keephills region near Lake Wabamun. There were many good reasons to expand its developments here. The company had already constructed ten units in the area, roads and transmission were largely in place, the North Saskatchewan River could supply water for cooling, and coal was readily available at the nearby Highvale mine. However, the coal seams ran south under the hamlet of Keephills. In order to proceed, it would be necessary to relocate the community.

The project design called for two 375 MW units, with room for two more units later. It had a large footprint: 2,500 hectares for the power plants and related facilities, and 6,200 hectares for the mine. TransAlta commissioned Keephills I in 1983, and the second unit a year later. During the construction of the first two Keephills generating units, the company applied for approvals for the construction of units 3 and 4, for commissioning in 1985 and 1986. Alberta's Energy Resources Conservation Board (ERCB) also considered a similar proposal from EPCOR for its Genesee I and II units. Wanting to encourage competition, the ERCB approved EPCOR's bid and delayed acceptance of the TransAlta project. Thus, though a disappointment to the company at the time, the severe economic downturn created a glut of power on the market, and TransAlta actually benefitted from the regulatory agency's decision.

Keephills 3 was not built until two decades later, a reflection of the depressed market conditions of the 1980s and the emergence of gas-fired units in the 1990s. But TransAlta's ability to add eight new units to its operational base between 1970 and the early 1980s was impressive. In an era of cheap coal, pulverized coal technology, and rapid market growth, the utility took advantage of its opportunities.

Keeping the Power Flowing

Each new power plant required transmission to get the electricity to market. In Alberta, the main power plants are west of Edmonton, but markets are spread throughout the province. TransAlta had built up an extensive transmission system since the 1920s that serviced its needs and connected other power generators to markets. During the 1970s and 1980s, however, the company had faced increasing opposition from landowners over transmission rights-of-way. Some people did not want the lines to go through their land; others demanded high fees. When TransAlta began plans for a new 240 kV line from Calgary to Lethbridge, tensions rose. Though the line ran through open country south of Calgary, a number of farmers and ranchers were determined to stop the project. Tough battles at the regulatory hearings were followed by three years of appeals through the courts. When TransAlta finally got approval for the project and surveyors began working on the right-of-way, one of them was physically assaulted by angry local residents.

Security and protection for employees came to a head in January 1981 when a bomb destroyed a steel transmission tower near High River. All employees were placed on full alert. A *Calgary Sun* headline read, "Bombing Onslaught Feared," and the RCMP speculated that further attacks could be coming.

Bombed tower near High River, Alberta, 1981. TRANSALTA COLLECTION

The company moved quickly to protect its employees and its assets and posted a $10,000 reward for information leading to the arrest and conviction of those responsible. TransAlta's official statement purposely downplayed the seriousness of the event in order to ease tensions. It was termed a "senseless act by someone who appears to harbor a grudge against the company." The province was more aggressive, and warned of a life sentence for anyone convicted.

Through these tensions, the employees showed remarkable calm and continued their normal duties. They were determined that the bomber(s) feel no satisfaction in this attempt to terrorize the company. TransAlta increased security and the bombing stopped. Nevertheless, transmission lines through remote areas were almost impossible to protect, and any downed live wires posed a threat of injury to employees or the public.

The dilemma for TransAlta was that new transmission was required as the company grew. There seemed to be no solution to deal with landowners opposed to the expansion. It might have been a new age in public attitudes, but, as one employee put it, "It was the end of the age of innocence for the company."

Interprovincial transmission raised a series of policy issues for TransAlta. The company had consistently opposed a national transmission grid, fearing it would turn into a government-subsidized way for utility companies in adjacent provinces to dump surplus power into the Alberta market. There was also very limited opportunity for any export of power from Alberta, with no link to the Montana system and only a tiny one to Saskatchewan. British Columbia presented the only opportunity.

In 1986, a major new line was commissioned through the Rocky Mountains to BC Hydro. This opened a new era of sharing power for mutual benefit through a 500 kV line. Alberta sold surplus coal-fired power to British Columbia at night, allowing its western neighbour to preserve its water-powered hydro capacity for daytime use. Then, at peak times, stored water generated electricity and came back through the transmission line to Alberta. It improved the reliability of both electric systems, and, equally important, the new transmission meant that power was available in emergency situations for either system. Additionally, it allowed for export of power into the western United States when there was excess capacity.

TransAlta also initiated collaboration with Washington State in the early 1990s. In 1991, the Alberta company signed its first firm export deal with the American hydro giant, Bonneville Power Administration. Terms of the deal stipulated that TransAlta supply BPA with $25 million worth of power over four years, to be delivered through TransAlta's link with BC Hydro. Market prices delivered a good rate of return for TransAlta.

The weather was less predictable. A spring blizzard in May 1986 brought down twenty-two steel transmission towers as well as three thousand wooden poles, disrupting service. Restoring power to customers demanded round-the-clock work by TransAlta crews—no small task during a blizzard. While power was restored as quickly as possible, the damage to the system took time to repair. The company had to mobilize outside crews and equipment and purchase poles from other provinces. It took the entire summer to replace and repair all the damaged lines and cost TransAlta more than $13 million.

There were also more regular problems, such as transformer fires from road salt, moisture, or other causes. In the 1980s, the company also began replacing the coolant in its transformers, as it contained polychlorinated biphenyls (PCBs)—highly controversial,

TransAlta completed a 500 kV transmission line from the Langdon substation southeast of Calgary to interconnect with British Columbia Hydro at Phillips Pass in the Crowsnest Pass area of the Rocky Mountains, c. 1986.
TRANSALTA COLLECTION

The new system control centre, 1985. TRANSALTA COLLECTION

toxic compounds regulated by government. The company now uses other coolants in its operations.

In 1985, TransAlta took an important step to integrate and improve the efficiency of the command structure for the hydro and coal power flowing into the grid. The company moved the hydro remote control centre from Seebe into a new state-of-the-art system control centre (SCC) in Calgary. This beating heart controlled much of the Alberta electricity system. The new facility managed the flow from the thirteen hydro and three coal-fired generating units along the transmission corridors to the local substations for distribution. A large map of the province with lighted lines and flashing lights covered a wall at the SCC. Five six-person crews staffed the centre around the clock. According to company management, this type of operational analysis was needed to maintain quality of service standards in an interconnected system that was becoming increasingly complex.

A Commitment to Sustainability

Born in the 1960s, the ecological movement became a major public and political issue in the 1970s, only to intensify in the 1980s. As scientific understanding of complex environmental issues grew, public concern increased, and policy response became more complicated and costly for industry.

While some businesses considered the movement a passing fad, more progressive corporations took a proactive approach. At TransAlta, Marshall Williams championed environmental stewardship as prudent business management and good community relations. TransAlta's commitment to change involved capital and operating costs, but management believed they were an important investment in the company's future. In the late 1970s, environmental initiatives became part of Calgary Power's capital budgets and its working code of ethics. As of 1981, TransAlta had installed environmental equipment worth $178.3 million—7.7 percent of total corporate assets. As environmental and emission abatement became a central part of corporate planning, funds were spent on engineering design, abatement technology, and operational systems.

Alberta Environment awarded TransAlta its first reclamation certificate in 1982 for 97.4 hectares of land at the Whitewood Mine, the second such certificate in Alberta mining history. At this time, the provincial government set a new standard for land reclamation, requiring companies to "replace existing topsoil and growth material" to a depth of 1.7 metres—up from the old standard of 0.45 metres. TransAlta saw that this new policy would dramatically increase the cost of reclamation. At the Highvale mine, for example, costs would rise to $42,000 per hectare—about a twenty-fold jump. In the end, TransAlta and the Alberta government worked together on field reclamation trials for some years, and, in 1986, the company was successful in demonstrating to Alberta Environment that only 0.55 metres of soil salvage and replacement was necessary. By 1986, TransAlta had reclaimed 971 hectares of land, which was returned to agricultural production or into wildlife reserves.

TransAlta environmental policy encouraged its employees to find innovative ways to convert wastes—a cost—into a product—a source of revenue—wherever possible. For example, for decades the company had been capturing fly ash—a by-product of burning coal—using first mechanical and then electrostatic precipitators on the stacks. The ash was then buried back into the mine. However, it was discovered that fly ash can be used as a replacement for 25 percent of the cement powder in concrete, adding strength and making the final product smoother. By replacing this portion of cement, the use of fly ash also reduces emissions from cement plants. In 1983, TransAlta Fly Ash Ltd. created a joint venture with Canada Cement Lafarge Ltd. to market fly ash.

In the early 1980s, TransAlta tested an innovative technique that promised to further reduce plant emissions and to help increase oilfield production in oilfields near generating plants. It was the first project of this nature in Canada and, decades later, was to become an important tool in the global quest to reduce greenhouse gases. TransAlta partnered with Dome Petroleum to use an amine solution to capture flue gas (CO_2) at the Sundance generating plant. Through a recovery process similar to water injection (which the oil patch has employed since the 1940s), the CO_2 process would extract oil that is unrecoverable through conventional means, adding to Alberta's total energy supply. Unfortunately, though the CO_2 scrubbing system was technically possible, it proved too costly at the time. As of its centennial, TransAlta was pursuing a new retrofit technology as part of its carbon capture and storage (CCS) with Project Pioneer at Keephills.

TransAlta undertook several other advanced research projects to address environmental issues. In partnership with the University of Calgary, TransAlta embarked on a test facility using wind power to pump water into hydro storage reservoirs where it could be used to generate power at a later time. Another project, with the Alberta Research Council, studied coal gasification as a way to greatly reduce emissions from coal combustion.

Not all solutions to environmental challenges were on such a large scale. TransAlta's John Tapics remembers a landowner near the Sundance plant who strongly objected to the use of herbicides to control vegetation under power lines. After brainstorming with the crew, the company came up with an innovative solution. It contracted the Saskatchewan Goat Company to bring in three hundred goats that ate all the plant life down to ground level under the watchful eye of two shepherds.

By the 1980s, wind power was emerging as an alternate form of energy in response to a demand for greener power. The Alberta government passed the Small Power Research and Development Act, a program that provided a market for wind energy by requiring utilities

continued on page 118

POWERING GENERATIONS

View of the cooling ponds at Sundance, 1988. TRANSALTA COLLECTION

Sundance plant at Lake Wabamun, Alberta, 1986. TRANSALTA COLLECTION

Whitewood mine, 1986. Beekeeping and the creation of wildlife reserves were part of the reclamation process. TRANSALTA COLLECTION

Reclaimed land at Whitewood mine site. Reclamation began in 1986. TRANSALTA COLLECTION

The Small Power Research and Development Act

Given Alberta's vast supplies of coal and petroleum, it is not always easy for innovative new systems to enter the power market. In 1984, the Energy Resources Conservation Board decided that private producers of wind energy could add their excess production of electricity into the grid and withdraw power from the system when their turbines were not operating, giving them a constant supply of electricity for their customers.

Entrepreneurs in 1985 formed the Small Power Producers' Association of Alberta and lobbied for better access to markets, accusing the larger companies of unfair market domination and only paying 1.6 cents per kilowatt for the power it accepted into its grid. For its part, TransAlta argued that the payment it made to windmill operators was fair, because it was the same as what it cost the utility to produce electricity.

Then, in 1988, the Alberta government passed the Small Power Research and Development Act to encourage entrepreneurs to branch out and thus to help diversify Alberta's power generation system. The act required existing utility companies to purchase power from the small power producers at a regulated rate, based on the replacement cost of a similar facility.

Ross and John Keating saw an opportunity. They formed Canadian Hydro, which, as a small producer, benefitted from the Small Power Research and Development Act. The company built its first power plant in 1990 and averaged a new hydro project each year that decade. In 1999, it purchased the Cowley Ridge wind plant near Pincher Creek in southwestern Alberta, the first commercial wind farm in Canada. Canadian Hydro added more capacity during the next decade, in Alberta, British Columbia, and Ontario, eventually buying out Canadian Renewable Energy Corporation, Vector Wind Energy, and GW Power Corporation, and became part of TransAlta in 2009.

Soderglen wind farm, Alberta, 2007. TRANSALTA COLLECTION

to purchase electricity from the small producers—at double the price of TransAlta's coal and hydro power. TransAlta was not opposed to wind power in principle, but felt that the government's decision was a costly and arbitrary change affecting the competitive marketplace. Later, between 2000 and 2002, TransAlta purchased all the shares of Vision Quest Windelectric in order to become a major player in wind energy. TransAlta expanded its investment in this same renewable energy source in 2009 when it purchased Canadian Hydro Developers. And this commitment to renewables continued to grow; as of the end of 2010, TransAlta was the first utility company to own and operate more than 1100 MW of installed wind capacity in Canada—24 percent of the country's total.

Though largely reliant on hydro and coal as fuels for generating electricity in the 1980s, TransAlta considered yet other alternatives, including nuclear power generation. Company president Bert Howard served as president of the Canadian Nuclear Association from June 1976 to June 1977. Nuclear power generation boasts low-cost fuels, but its capital costs are high, and plants in Ontario, for example, had suffered serious cost overruns. Provincial government there guaranteed the debt, but such a backstop would not be available to an investor-owned utility in Alberta. Siting was also a concern in central or southern Alberta, given the large volumes of water required for cooling. And nuclear power generation faced opposition from the public, so TransAlta shelved the nuclear option.

Regulatory Challenges

For much of its early history, TransAlta operated in a lightly regulated sector, but by the 1980s an increasingly complex world had created a need for more government oversight of industrial activity in Alberta. For example, the company had avoided rate hearings for decades because the provincial PUB had no jurisdiction over rates unless they rose. Calgary Power's economies of scale until the 1960s meant that the price of power decreased each year, even as the cost of living rose, so the company considered its rates reasonable.

The first challenge to rates had taken place in 1961, in Red Deer and Jasper Place (now part of Edmonton), when two municipal wholesale customers registered a complaint with the PUB over the price Calgary Power was charging for electricity. Lawyers were busy for the remainder of the 1960s, sorting out complex legal issues. The PUB called upon Calgary Power to justify its rates, and so the company compiled a detailed report of its expenses back to 1930, when the province achieved control over its natural resources. According to Harry Schaefer, the company's chief financial officer at the time, Calgary Power's return on capitalization amounted to a modest 6.28 percent, while other utilities at the time were reporting a 7.5 to 10 percent return.

As a result of this hearing, the company applied to have its water power licences amended to fall in line with other utilities in the province—some had been under federal control—and in September 1972 the licences were changed by mutual agreement between the province and Calgary Power. It then began preparing for its first rate hearing before Alberta officials.

Once the company entered the 1970s, an era when electricity rates were no longer falling, regulatory hearings became a necessary part of the public process. The company had to assure customers and investors that utility rates were fair before committing capital to projects. If investors were not adequately rewarded, they would not provide the capital needed to build new plants, and customers would not have a reliable supply of power. And Calgary Power had to have confidence that its rates would provide a decent return on the capital being invested. The PUB's role was to adjudicate the process and to set fair and reasonable rates to protect the interests of all parties. The process in Alberta is generally considered to be equitable. Usually, if everyone is a bit unhappy, it means that the right balance has been struck!

One of the most difficult regulatory issues the company ever faced resulted from the creation of the Alberta Electric Energy Marketing Agency (EEMA), which operated from September 1982 through 1995. The provincial government created this agency in response to complaints from other electrical generating companies in the province who claimed that TransAlta's large market share—about 80 percent—provided unfair advantages to the company. They claimed that the utility's low prices were a result of its market dominance. The customers of ATCO and Edmonton Power were at a disadvantage, they argued, having to pay higher prices in this non-competitive market. TransAlta countered that it had carefully built a low-cost system and that its customers and shareholders should not be penalized for this efficiency.

According to TransAlta's 1982 Annual Report, the purpose of EEMA was to "facilitate equalization of electric power generating and transmission costs throughout the province." All generators would be required to transfer their output into a common pool, and electricity would be marketed at a uniform price set by the PUB. In TransAlta's view, the company would be selling its electricity to EEMA and then buying it back at higher prices, in effect subsidizing the higher costs of its provincial competitors. The company opposed this new agency, seeing in it the destruction of its corporate model of integration of generation and transmission. Moreover, the agency policy effectively levied a new tax on TransAlta customers. The provincial government promised to subsidize the equalization for the first five years of the program, but TransAlta concluded, "As the subsidy is withdrawn it seems inevitable that our customers will pay more than they would pay without the Agency."

In an environment where all power was dispatched through the government pool, the economic cost to the company, its customers, and its shareholders grew. The increase, as much as 14 percent on a typical electric bill, amounted to hundreds of millions of dollars over the next fifteen years. As a result, the company lost its low-cost competitive advantage. EEMA diverted value from the company and helped its rivals expand, changing the structure of the electric utility business in Alberta.

Diversifying the Company

During the downturn of the 1980s, in an economic climate where there was no room for growth as a generator of power, TransAlta diversified its operations beyond electricity. In 1981, it created TransAlta Resources, a subsidiary whose mandate was to develop non-utility assets, with Harry Schaefer as its leader. This strategic move expanded the company's operations into oil and gas, technology, and management services. TransAlta also purchased common shares and debt in Canada Northwest Energy, with oil and gas holdings in Alberta, Saskatchewan, the U.S., and offshore Spain. Northwest also held stock in Panarctic Oils, a frontier player that was drilling in the High Arctic. Investments of nearly $100 million in the 1980s made TransAlta the largest shareholder, holding 40 percent of the common shares by 1987. But, as the world price of oil plummeted, the company's investment in Northwest Energy lost value and TransAlta sold its shares in 1990.

TransAlta Resources also invested in a 50 percent stake in Alberta Energy Corporation's AEC Power, a small utility that provided electricity and steam to oil sands facilities being built by Syncrude in the late 1970s. This move into cogeneration foreshadowed future developments and allowed the company to gain entry into the rapidly expanding oil sands sector. This new "behind the fence" application utilized a gas-fired generator and moved the fuel to the market, rather than the electricity. For example, highly efficient natural gas first fired the electrical generating unit and then employed heat from the process as part of the method for removing bitumen from the oil sands.

TransAlta entered the computer business in the mid-1980s, purchasing 50 percent of the stock in Keyword Office Technologies of Calgary for $11 million. It seemed a promising business in an era when clients needed software capable of exchanging documents between operating systems. After some initial success, the company expanded into other lines. But, overwhelmed by fast-moving competitors, like many other promising early computer companies, TransAlta eventually sold Keyword in 1995.

TransAlta Energy Systems was another subsidiary created during this period. This company marketed energy conservation, fire protection, and security service for buildings. It also investigated combustion technologies for improving environmental performance, and in 1986 it acquired the world rights from Rockwell International for a new combustion technology to lower nitrogen oxides (NO_x) and sulphur dioxide (SO_2) emissions. This research and development project showed promise in the fight against acid rain, but development costs proved too high and the project was abandoned.

Of the various diversification projects, only the Syncrude project proved to be a solid long-term business opportunity. However, these experiences encouraged a sharpening of the company's focus as it evolved during these challenging times.

In the 1980s and early 1990s, company presidents Marshall Williams and Ken McCready encouraged programs to enhance productivity and efficiency. They knew that people were the key to TransAlta's success. In 1981, the company created a Productivity Council with representation from managers and employees. The council met monthly and its Action Recommended Program considered hundreds of submissions. By 1988, the council had reviewed almost five thousand recommendations, 1,047 of which were formally implemented. Employees whose ideas were implemented received ten TransAlta common shares. These innovative ideas saved the company millions of dollars, improved productivity, and increased customer goodwill. The results were evident in the Canadian Electricity Association rankings, where six of the ten most productive thermal units in Canada were TransAlta's. This best-in-class performance was a great tribute to the employees involved and to the overall management system.

Part of the reason for this high performance was the human resource department's skills training and education programs. The company worked hard to make sure that each employee had the skills and experience to feel comfortable on the job. Mentoring helped, as did formal training. Apprenticeship and journeyman instruction were provided at the company's Red Deer Trade Training School, while others attended the institutes of technology in Edmonton and Calgary. Extra training and cross-training allowed for flexibility in the plants, because each worker could do several jobs.

Generation and Energy Marketing: A New Approach

By the 1990s, despite attempts to diversify assets, TransAlta's profits remained static and Board Chairman Marshall Williams and President Ken McCready saw little room for growth in the Alberta utility market. The company began to explore profitable new opportunities that built on TransAlta's strengths and its excellent Alberta experience.

NET INCOME OF TRANSALTA, 1985–90

Year	Net Income
1985	$180.9 million
1986	$187.1 million
1987	$176.6 million
1988	$173.9 million
1989	$150.0 million
1990	$145.3 million

Ken McCready

Ken McCready, 1988. TRANSALTA COLLECTION

Visionary, resourceful, a great fan of innovation and thoughtful reflection, Ken McCready rose through the company's ranks to become its CEO in the tumultuous period from 1985 to 1996.

Hired by Hap Hansen, he started with Calgary Power in 1963 and immediately became interested in the economic side of the company. Ken's interest in economics made him fit for the work of a rate engineer. Though Calgary Power had purchased its first computer in the early 1950s, McCready and Harry Schaefer purchased early Macintosh computers with their own funds and used them for planning and data manipulation purposes at Calgary Power.

Economies of scale and the incredibly low price of coal fuel had allowed Calgary Power to consistently lower its rates from the 1950s through the early 1970s. But the forces of inflation had caught up by 1972, eroding the economic benefits of the boom, and necessitating Calgary Power's first rate increase. Rate hearings became onerous during the 1970s as more interveners challenged the company's operations. The fact that Calgary Power had always been the lowest cost provider made for tension in Alberta—consumers who bought power from other suppliers wanted it as cheap. In 1982, the Lougheed government implemented a pragmatic political solution with the Electrical Energy Marketing Agency (EEMA). The goal of EEMA was to equalize the rate to consumers throughout Alberta. The rate for Calgary Power's customers went up in order to subsidize electrical bills in other parts of the province.

Environmental issues were also important during Ken McCready's tenure. By late 1989, 21 percent of Canadians identified the environment as a key concern—up from just 5 percent earlier that same year. In 1990, the Alberta government created a twenty-six-member Roundtable on the Environment and Economy—including three environmental advocates—to advise it on ways to develop environmentally sound economic policies. Ken, as one of the founders of the hugely influential World Business Council for Sustainable Development and a respected thinker on sustainable development issues, was appointed to chair the panel. Ken saw that internalization of the costs of carbon was vital to TransAlta and to industry throughout Canada.

During Ken's time as leader, the company ranked well internationally as a power producer and in the middle of the pack on transmission and distribution. Ken understands the reasons TransAlta retrenched to its current position as just a power producer, saying, "The best opportunities to grow the company for the shareholders and the employees was to move on your strength, build on your strength, and the strength was generation." Still, Ken thinks part of the corporate culture was lost during the transition. He believes the future of TransAlta continues to be dependent on technological developments, innovative thinking, and expansion into alternative forms of energy production. "It's still magic—you turn a switch and the light comes on." Ken McCready passed away on July 30, 2011, in Calgary.

TransAlta began the search for new technology, new markets, and a new fuel option to complement hydro and coal. Natural gas-fired electrical generating facilities were coming into their own, opening the way for the development of smaller independent power plants (IPPs) to service local markets. Low natural gas prices allowed jet engines, with their power and efficiency, to be adapted as power generation turbines. Natural gas could be moved easily and cheaply by pipeline to any site, and the turbines could be fired up as required by consumer demand. The Achilles heel, however, was the volatile price of natural gas.

Guided by its experience with local electrical generation at Syncrude, TransAlta moved aggressively into this new sector. In the early 1990s, it explored projects in British Columbia, Alberta, Ontario, and Quebec. In 1990, the company signed contracts in Ontario that resulted in "greater energy efficiency and reduced effect on the environment." A $100 million, 110 MW cogeneration plant at the McDonnell Douglas's airplane wing assembly plant in Mississauga supplied power, heat, compressed air, and purified water to the plant and electricity to Ontario Hydro. A smaller plant in Ottawa served the Health Sciences Centre—a group of hospitals and related facilities. When an ice storm in 1998 cut out the centralized electrical supply, this plant provided electricity for these essential services. A third independent power plant provided power for the Chrysler manufacturing complex in Windsor, Ontario. This strategic shift to natural gas and into limited partnerships in the Ontario market allowed for another step in TransAlta's evolution.

The new strategic vision also included international operations. In 1993, TransAlta invested $100 million in two operating hydro units and two under construction in the Limay River Valley in western Argentina. TransAlta was impressed with the market potential: electricity consumption was rising faster in Argentina than in Alberta. With Duke Energy (U.S.) and Chilgener (Chile), the company brought a 59 percent interest in the 1400 MW project, designed to supply 10 percent of Argentina's power. This daring move was reminiscent of Max Aitken's forays into the Latin American utility market in the early 1900s.

TransAlta also looked halfway around the world to invest in a cogeneration plant and marketing operations in New Zealand. It then considered opportunities in Australia, developing a circum-Pacific strategy. The company was also pursuing projects in China and India. In five years, the quest for international opportunities had resulted in promising projects in ten countries.

It was a risky strategy for an Alberta-based company with limited experience in the international utility business. The rapidly changing geographic diversification challenged TransAlta's human resources as well as the mindset of its investors—a group of people who valued the stability of a conservative, reliable, dividend-paying stock. Thus TransAlta developed a formal set of criteria to apply to every international project. The project had to align with the core business of the company; it had to have alliances or active partners in the local market; the country had to have a stable political environment as well as a stable currency and sound financial institutions; and it had to serve a growing economy with rising electricity demands.

Though some American companies also pursued an international strategy, TransAlta's vision was unique among Canadian electrical utilities. Unfortunately, the political and economic risks associated with developing countries were not fully appreciated and, as a result, the company eventually backed out of several of its international initiatives. Nevertheless, the "tolerance for learning" that Marshall Williams referred to in his 1988 speech was an invaluable and positive gain for management. Throughout this difficult period, the company successfully defended its market share and balance sheet integrity. It also expanded into new ways of generating electricity and began preparations for more competition in the less-regulated economic climate that was to come.

RECESSION AND MARKET TURBULENCE 123

"At a time of such change, unique opportunities become available to those who are prepared." HARRY SCHAEFER AND WALTER SAPONJA, 1995

CHAPTER SEVEN

Preparing for the Future

TransAlta's story is the tale of a company that provides a reliable product to its customers, a predictable return to its investors, and serves the needs of society. It has faced its challenges and made its mistakes. Sometimes it has followed when it could have taken the lead, but each time the people at TransAlta have worked together to sharpen their focus, reconnect with the community, and chart a course for the future. As the largest investor-owned electricity generator and trader in Canada celebrates its centennial, its people can take pride in how far the company has come. What once was a small operation on Alberta's Bow River has evolved into an international player: smart, competitive, and sustainable.

Over the past century, the company has had to recreate itself several times, changing and evolving as the situation demanded. This has called for skill, innovation, determination, and commitment—qualities that were essential in the period covered in this chapter. By the mid-1990s, a perfect storm was forming on the horizon: TransAlta's strategic planning was at a crossroads, its leadership was in transition, earnings were flat, and Alberta's utility sector was undergoing a fundamental overhaul as it moved in fits and starts to a deregulated and more competitive business environment. By 1995, it was obvious that the company had to reinvent itself yet again.

Fundamental change is now impacting our industry on a global scale. A new more competitive market is emerging. At

Kent Hills wind farm, New Brunswick, 2008. TRANSALTA COLLECTION a time of such change, unique opportunities become available to those who are prepared. TransAlta has been positioning for the change by improving performance in its core business. At the same time, we have been gaining experience and building a presence in regions where industry change provides opportunities to grow long-term shareholder value. We are also acquiring and developing the expertise which will strengthen our competitiveness.

– Board Chairman, Harry Schaefer, and President and CEO, Walter Saponja, 1995 Annual Report

These are galvanizing words indeed, coming from two long-time TransAlta employees who had risen to lead the board of directors and the management team in the 1990s. During this critical time of transition, their leadership, as well as that of the board, allowed the company to prepare itself for a new business climate and embark on the most rigorous changes in governance, management, and culture in the company's one hundred-year history.

Asking the Tough Questions

Harry Schaefer had served as TransAlta's chief financial officer from 1975 to 1993 and had raised significant amounts of capital for the company's expansion during boom times and hard times, reassuring nervous investors and bringing financial discipline to the corporation. As a result, he learned that good governance and good relations with investors were key to selling equity and debt instruments.

By the early 1990s, it had become increasingly apparent to the general public that not all corporations were guided by appropriate boards of directors. Obviously, any investor looking to entrust large sums of capital with TransAlta wanted assurance that its leaders were competent, skilled, and ethical. Good governance was good business.

Following on the heels of a number of spectacular financial failures in the business community in Canada and around the world, the Toronto Stock Exchange commissioned a study into the health of Canadian corporations. In late 1994, it published the Dey Report, titled *Where Were the Directors? Guidelines for Improved Corporate Governance in Canada*. The report demanded public disclosure of the relationship between boards and management, as well as an answer to the question, "Where were

TransAlta's geographic focus, 1995. TRANSALTA COLLECTION

the directors?" when the decisions that led to catastrophe were being made. As part of that project, the researchers contacted TransAlta, and Schaefer provided them with up-to-date information on the workings of the company's board.

"To do the job effectively, directors need to understand the workings of the business and provide their support, strategic insights, and expertise to the CEO," Schaefer wrote. "Equally, they need to be able to look the CEO in the eye and ask the tough questions that need to be asked."

The Dey Report recommended that "the board of directors of every corporation should be constituted with a majority of individuals who qualify as unrelated directors." These individuals should be "free from any interest and any business or other relationship which could, or could reasonably be perceived to, materially interfere with the director's ability to act with a view to the best interests of the corporation …"

The report stated that other hallmarks of good governance are a strategic planning process, a committee to recruit and nominate board members, and a separation of the role of chairman of the board from the CEO role. Well-governed companies should also have a risk-management strategy, a method for monitoring senior management, and a succession plan.

A review of the membership of TransAlta's board dating back to 1981 shows that until 1990 about two-thirds of the directors were independent of management. During the 1990s, that ratio rose to 80 percent, and since 1997 fully 90 percent or more of board members have been independent directors. In its 1995 review of governance issues, the company reaffirmed that the board would "support and encourage" management in their duties, and management would make "wise use of the board's skills before decisions are made on key issues." The company's current guidelines vary little from the ones published in 1995, and it has won several national awards for corporate governance over the years.

In the mid-1990s, for the first time ever, a TransAlta board meeting was held without the president and CEO of the company in attendance. "The value of non-executive meetings was immediately evident, and they were continued from that time forward," Schaefer recalled.

Leadership in a New Era

A new leader, with a new set of skills, arriving in the face of fresh opportunities afforded by fundamental changes, can invigorate and rejuvenate a company. In early 1996, the TransAlta Corporation board and its president and CEO, Ken McCready, concluded it was time to introduce new leadership to see the corporation through the major changes occurring in the industry and to prepare the way for a post-regulation business climate. Deregulation of the electrical utility industry meant that the TransAlta board needed to find a leader with extensive experience in the unregulated business world. For the first time in the company's history, this leader would be chosen from outside the close-knit corporate culture.

In the interim, the board entrusted the leadership of the company to Walter Saponja. During his thirty-seven-year career with TransAlta, he had held progressively responsible positions, as senior vice president for operations, then president and chief operating officer of TransAlta Utilities, and then president and chief executive officer for the entire group, TransAlta Corporation, in 1996. Saponja had been involved with the expansion of the company's thermal power generation facilities for three decades, including design, construction, commissioning, and operations. He had an amazing ability to comprehend technical detail and its significance for the reliability of power plants.

The challenges facing Saponja as interim leader were many and varied. Existing regulatory systems were still in place, and governments were moving steadily toward a deregulated utilities climate, changing or eliminating some programs and creating others, all with

continued on page 130

 POWERING GENERATIONS

Three Generations of Schaefers at TransAlta

Harry Schaefer, who has served TransAlta in many ways, is sandwiched between two TransAlta leaders of two generations. His father, Ernie Schaefer, was an engineer with Calgary Power, hired on in 1929. A practical person, he could build just about anything. Harry's son, Rob Schaefer, is part of the management team as of 2011—the year TransAlta celebrated a century of providing electricity.

As a boy, Harry had to make a choice between becoming a forest ranger or a chartered accountant. "In the end, mosquitoes were the determining factor," he recalls, and he progressed from a paper route to the University of Alberta, where he received a degree in commerce.

Married, and with a first child on the way, Harry accepted a job with the accounting firm KPMG in Montreal, where job opportunities seemed greater than in Calgary. But when Calgary Power offered him the position of assistant controller in 1963, he jumped at the chance to move back to Alberta and work for the company his father had served for decades.

Harry eventually became chief financial officer from 1975 to 1993 and developed several programs to make the organization attractive to investors, raising some $4 billion in capital to finance the company's growth. In 1981, he became president of non-regulated operations and joined the board of directors. He became board chair in 1991.

Since retiring from TransAlta in 1996, Harry has been involved with a number of public boards, including TransAlta Power LP from 1998 to 2000. He also serves as a business adviser to various institutions and has contributed to the educational programs at the Institute of Corporate Directors, passing on his knowledge on mergers and acquisitions in more than forty-five workshop sessions over the years.

Rob Schaefer, Harry's son and Ernie's grandson, is a chartered accountant with expertise in financial and project development in the energy field. He joined TransAlta in 2008 in business development and is now TransAlta's vice-president of Commercial Operations and Development. In this role he is responsible for business and corporate development and growing TransAlta's customer base, commercial management, and trading operations.

Harry Schaefer, c. 1980. TRANSALTA COLLECTION

Ernie Schaefer was involved in transforming an old aircraft hanger into Calgary Power's office building, pictured here in the 1950s. VIC JONES

"Mr. TransAlta"

Born and raised in Alberta, Walter Saponja brought a strong work ethic to the job when he hired on with Calgary Power in 1961, after graduating with a mechanical engineering degree from the University of Alberta. He was always involved with athletics and considered teamwork an important part of success.

Like many other company people, Saponja was attached to Montreal Engineering for a year and a half during the construction of Wabamun Unit 3. He was impressed with the talented employees of the Montreal firm and learned a lot from them, finding them humble and unpretentious. He developed what he called a "secret elixir," which was simply to "do the best job you can do for what you've been asked to do and the rest will look after itself." He frequently passed this advice on to ambitious young employees who sought his counsel.

Service to the customer was paramount. Company president Marshall Williams used to "preach to me that if we keep our costs to the customer low, the shareholder will end up being looked after in the long haul." It was the mantra for Calgary Power and TransAlta for decades. The company was a storehouse of committed employees, in all areas, who prided themselves on being the best at what they did.

Walter considers his greatest achievement with the company was being part of an extremely efficient operation, and, in particular, he believes that "nowhere were there other coal thermal power plants that operated with as few people as we did."

Walter was involved with the downsizing of staff by about 15 percent in 1993-94, a first in the company's history—and "it did not go over very well." In response, he reached out to the staff. "I got our human resources people to organize what we called coffee pot sessions—casual sessions where you'd get twenty people around several cups of coffee for an hour and answer questions and talk about corporate strategy."

The decision to focus the company into a core business—generation—was fundamentally sound. "Operational excellence" and competitiveness were the driving forces during Walter's time at the company and remain pivotal at its centennial.

In 1999, Walter received a Distinguished Service Award from the Canadian Electricity Association in recognition of his contribution to the electric utility industry.

Walter Saponja, 1997. TRANSALTA COLLECTION

Steve Snyder

Steve Snyder was born in Montreal and holds a Bachelor of Science in Chemical Engineering from Queen's University and a Master of Business Administration from the University of Western Ontario.

Prior to joining TransAlta, he was president and CEO of Camco Inc, GE Canada Inc., and Noma Industries. As head of Noma Industries from 1992 to 1996, Steve advanced the company from a largely Canadian consumer products-manufacturing company to a North American industrial products company. As chairman and CEO of GE Canada Inc., Steve continued with the transformation of GE's Canadian-based businesses into global competitors. Steve is now president and CEO of TransAlta Corporation as well as a director of TransAlta Corporation and Intact Financial Corporation. In 2005, he was awarded the Alberta Centennial medal, and in 2010, the Energy Council of Canada named him Canadian Energy Person of the Year.

During his time at TransAlta, Steve has seen three large shifts in the electrical utility business. The first was deregulation. Up to that time, the company had a contract to sell its electricity in a regulated market, with low returns, but very low risk. Deregulation opened the door to competition and new markets.

The second upheaval was from about 2000 to 2005. "When the market tried to deregulate, the industry had visions of sugarplums," Steve recalled. Natural gas technology was seen as the solution to power generation challenges. Speculative traders got into the business. Managers who had excelled in a regulated environment struggled to adapt to market competition. Many companies expanded too quickly and made poor investment decisions. The result was many industry bankruptcies and public policy makers, concerned about price spikes and reliable supply, moved to regulate much of the sector. "This is a steady, long-cycle industry, and you need to have overcapacity for security of supply."

The third major shift is the one facing the industry currently—the carbon question. Though concern over greenhouse gases has been part of strategic planning at TransAlta since the 1980s, Steve said, "Dealing with the carbon issue means putting in all new technologies—from manufacture, to distribution, to end use." Steve expects massive technological change in the next decade. "The decisions that have to be made in the next ten years include a huge degree of uncertainty: uncertainty in the technology, uncertainty in the price of

Steve Snyder, 2008. TRANSALTA COLLECTION

carbon, uncertainty in the regulation, uncertainty in the targets. And yet, if we do not make decisions today, the results will not be there ten years from now."

Reflecting on TransAlta's history, Steve believes that it has held true to some central values: "A focus on people, a focus on cost, a focus on service, of being willing to take bold moves, but doing it—at the same time—with your eyes still on the ground.

"In 2050, the world's electricity is going to be totally different than it is today. I believe there will be distributed generation: people will have little mini-turbines or solar technology on their houses, neighbourhoods will be connected, there will plug-in depots.

"The challenge we face is to get from 2011 to 2050."

the intent of making way for the future. But the transition to the new system had been awkward and, in TransAlta's opinion, often unfair to the company and its customers.

For example, the Alberta Electric Energy Marketing Act (EEMA) created uniform electricity prices across the province. EEMA costs were phased in by the province; they started at tens of millions of dollars each year in the early 1980s, topped $117.5 million in 1990, and rose to $231 million in 1995, the last year of the program. From the company's point of view, under EEMA, TransAlta customers ended up subsidizing other Alberta consumers. Following the demise of EEMA, cost averaging of existing generation in the province was accomplished through Purchase Power Arrangements, or PPAs.

In May 1996, TransAlta's board elected a new chairman, Dick Haskayne, to succeed Harry Schaefer as he moved into retirement. Haskayne had served on the TransAlta board since 1991 and brought with him years of experience and expertise in board and governance issues. As a businessman who was not an employee of TransAlta, Haskayne brought an outsider's perspective to the shared objective of the directors—that of acting in the company's best interest. Haskayne had a good relationship with Walter Saponja, working closely with him for the remainder of his time as leader of TransAlta Corporation, and was able to oversee the transition to a new corporate culture.

In September 1996, Steve Snyder was appointed as the new president and CEO. According to Haskayne, Snyder "had a superb track record as an entrepreneurial leader focused on business development and marketing: two key areas that would shape TransAlta's future." Snyder, in turn, appreciated Haskayne's contribution to the board.

The new CEO knew that change was necessary. While he saw value in the previous administration's attempts to reduce risk by diversifying operations into other parts of the world, this strategy had not been a complete success. "The problem was when you spread it out to reduce your risk, you did reduce your risk, but you also reduced your profits," Snyder recalled.

Independent Power Projects

Companies innovate in order to survive and thrive, and, since the late 1980s, TransAlta had vigorously pursued non-regulated independent power projects (IPPs). IPPs and energy marketing were both non-regulated business opportunities that offered room to grow by expanding the company's diversification efforts and taking advantage of new opportunities.

Starting in 1983, TransAlta Energy Systems, a subsidiary, began exploring IPP investment opportunities. In 1988, the company was considering natural gas-fired peaking plants and cogeneration projects—a strategy that has become central to TransAlta's diversified fuel strategy. As a result, it landed two cogeneration projects in Ontario in 1990.

These projects burned natural gas—a relatively cheap fuel throughout the 1990s, though it had spiked along with oil in the

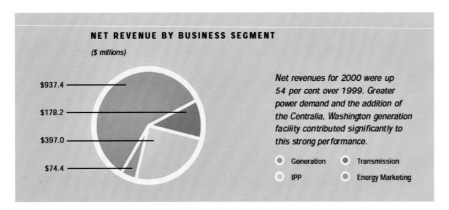

Deregulation happened quickly after 1995. TRANSALTA COLLECTION

early 1980s. As a by-product of the electricity generation process, the cogeneration plants also provided steam, which was sold to third-party customers for industrial purposes.

Natural gas-burning gas-turbine power plants could be located near the load centres—that is, near the customers. They were smaller, requiring about one-third of the time to build compared to a large coal-burning plant. They also required less capital, though their fuel costs were higher.

In a time of increasing environmental concerns and economic and regulatory uncertainty, cogeneration offers TransAlta opportunities for corporate growth. Taking advantage of our expertise in engineering, financing and operations, we plan to become a leader in Canada's cogeneration market.

–1990 Annual Report

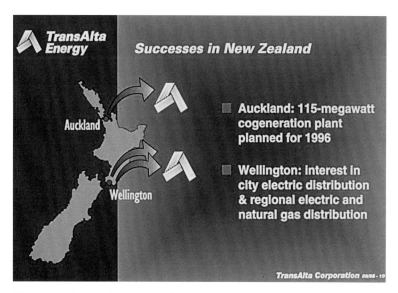

TransAlta plants in New Zealand, 1995. TRANSALTA COLLECTION

The first two cogeneration power projects, in Toronto and Ottawa—with Ontario Hydro as the power customer—required a $168-million investment and promised to provide lifetime revenues of almost $4 billion. These plants began operating in 1993 and performed well. They were available to produce power more than 90 percent of the time during their first year in operation, and rose to 96 percent in 1994 and 1995. TransAlta's cogeneration facilities were the first in the world to use the high-efficiency General Electric LM6000 gas turbines, whose technology derives from the jet engines used in aircraft.

In the mid-1990s, the company began construction of another cogeneration plant at Windsor, Ontario, and signed a long-term power purchase agreement with Ontario Hydro. Though the company's first independent power projects were cogeneration facilities, it also applied its technical expertise to installing gas-fired facilities that produced only electricity. And, in 1995, TransAlta started building a plant in Auckland, New Zealand—a joint venture with New Zealand's Mercury Energy Limited. The New Zealand operations proved to have excellent growth potential; TransAlta expanded its efforts there until it grew to be that country's largest retailer of power.

In 1997, the corporation set out to double its IPP business. TransAlta's independent power projects grew steadily over time, and by 2003, the generating capacity provided by the IPP plants had tripled to nearly 2500 MW.

Though coal, by comparison, was still producing most of the company's power in 2003—4777 MW—it had independent power projects in seven Canadian locations, including a cogeneration plant that provided power to Suncor Energy at the Alberta oil sands. There were similar projects in the United States, Mexico, and in Western Australia, making TransAlta a major player in the IPP industry.

The IPPs allowed TransAlta to expand its business activity beyond the regulated electricity sector and build generation capacity wherever it was required. If an economically stable country needed

continued on page 134

Energizing Communities with More than Electricity

For decades, a service truck with the Calgary Power logo on its side was a recognizable presence in the communities in which the company operated. Today, in a society where few people understand how the electricity they use in their homes is generated or transmitted, TransAlta's community presence takes a very different form. Although TransAlta sold off its distribution and retail segments in 2000, the one thing that has not changed is the belief that success comes from energized communities.

Throughout its history, TransAlta has strived to support the sustainability of its operating communities. The list of projects and organizations that have been supported by both employees and the company is lengthy. Currently, TransAlta's community investment strategy is to strengthen the communities where its employees live and work by focusing on four areas: arts and culture, education and leadership, environment, and health and human services.

Arts and Culture: TransAlta supports the cultural fabric of communities through initiatives that enrich the lives of employees and their families. The company's interest in supporting a community's culture includes a long association with one of Canada's largest tourism destinations—the Calgary Stampede. Calgary Power became formally involved with the Stampede in 1938 when it began sponsoring the G. A. Gaherty Trophy for the winner of the North American Bucking Horse Riding Contest. TransAlta's collaboration with the Stampede has grown steadily over the decades. The company helped take the Grandstand show to a new level of performance when it supported the building of a new stage, and went beyond this gesture with its support of the fireworks extravaganza, known as TransAlta Lights Up the Night, which can seen throughout the city. In 2007, the company pledged $2.5 million to support the development of young Albertans through the tuition-free Young Canadians School of Performing Arts, which has a long history with the Stampede.

The communities in and around Wabamun Lake have been important neighbours of TransAlta. The company's support of Edmonton's Fringe Theatre Adventures Society, dating from 2003, has resulted in the staging of the Edmonton International Fringe Theatre Festival, which brings together an international audience and drives the development of emerging artistic talent to enhance the city's arts community. In addition, TransAlta's support has transformed a small theatre into the TransAlta Arts Barns, a year-round community gathering place, and made possible the annual TransAlta Family Theatre Series, a season of family-focused and family-friendly plays.

Education: Educational initiatives train and develop existing and future TransAlta employees. TransAlta's educational initiatives include a multi-year partnership with the Southern Alberta Institute of Technology (now SAIT Polytechnic), which resulted in the creation of the TransAlta Electrical Power Industry Centre (epiCentre) in 2002. The TransAlta epiCentre provides instructional excellence in conventional and alternative sources of power generation. From 2003 to 2008, TransAlta's $4.5-million investment has provided scholarships, internships, and mining and plant equipment.

In 2007, TransAlta invested in the University of Calgary's Institute for Sustainable Energy, Environment, and Economy (ISEEE). This

Grandstand show, Calgary Stampede, 2010. TRANSALTA COLLECTION

cross-faculty institute aims to make the U of C Canada's top university in energy and environment studies. The ISEEE coordinates and facilitates research and training with the goal of improving the sustainability of Alberta's and Canada's energy systems. Projects combine natural, applied, and social sciences and bring together academia, industry, and government.

The Environment: TransAlta partners with organizations that help strengthen the company's corporate environmental initiatives through reclamation and conservation programs, air and water management studies, and student-focused programs and field trips. The company is also involved with grassroots initiatives that improve community-based sustainable land management. In 2000, the company became the lead corporate sponsor of the TransAlta Rainforest at the Calgary Zoo, which exemplifies the zoo's efforts to preserve plant and animal species and provide conservation education opportunities to over 1 million people annually.

In 2006, TransAlta donated $25,000 to the Trout Unlimited Water Conservation Fund, and, in 2010, with the stoppage of the Wabamun coal plant, the company contributed nineteen acres of critical shoreline habitat on Wabamun Lake to the Alberta Fish and Game Association's Wildlife Trust Fund. This property supports one of the largest breeding colonies of western grebes in Alberta. A significant number of other waterfowl also use the site, including eared and horned grebes, more than fourteen species of dabbling and diving ducks, and Canada geese. The Wildlife Trust Fund—the oldest habitat land trust in the province—will ensure the environmental and habitat values of the land are maintained for future generations.

Health and Human Services: TransAlta supports health, safety, and wellness programs in its operating communities and in neighbouring communities. It has also helped fund a Job Safety Skills program in more than one hundred Alberta high schools. The company contributes to the Family Resource Centre at the Calgary Urban Project Society, has sponsored a safety program for children in partnership with the RCMP called Safe Eyes and Ears, and has supported highway clean-ups and toxic waste round-ups.

In 2002, TransAlta contributed $1.5 million to the TransAlta Tri Leisure Centre in Edmonton. The centre includes two ice arenas, indoor soccer facilities, and a water park. That same year, TransAlta partnered with Champions Career and Employment Service Centre, an innovative, tri-sector partnership of non-profit disability organizations, government, and major Alberta corporations that assists persons with disabilities. In addition to funding, TransAlta provided employment opportunities and executive expertise. And, in 2003, TransAlta announced a multi-year donation to Hull Child and Family Services in support of a new centre to deliver its community services to high-need families in east Calgary.

TransAlta's annual United Way campaign is the most recent manifestation of the company's tradition of giving back to the community. In the 1930s, Calgary Power supported the Boy Scouts and the Red Cross, and it started giving modest amounts to the Community Chest and other charities in the mid-1940s—renamed the United Way in 1973. TransAlta's involvement with this important, volunteer-driven annual campaign remains strong, and the campaign's success is the result of employees' and retirees' efforts. The company's total donation to the United Way has passed the $1 million mark since 2000.

The TransAlta Rainforest at the Calgary Zoo. TRANSALTA COLLECTION

electricity to fuel its economic growth, TransAlta was interested in competing for the contract to supply the power.

"By 2000, the IPPs and the historic generation—once deregulation took place—were integrated into a single company," recalled Dawn Farrell, who served as the executive vice president of IPPs in 1998. "The company went from having an IPP division to really being a generation company with a diversity of assets, a diversity of fuels, and a diversity of markets."

Power Purchase Arrangements

As part of the provincial government's promise to deregulate the electrical industry in Alberta, it introduced the Electric Utilities Act in 1995. The objectives of the act were to increase competition, provide choice to consumers, and encourage new types of generation. In 1998, the province amended the act, simplifying the process for approval of generation and setting a level playing field for all merchants and consumers of electricity. The government also created the Alberta Power Pool in 1996 to handle the buying and selling of power in a more open market. The 1998 amendments allowed for a twenty-year transition period, during which the provincial government and the electrical utilities would work out any arising issues.

Looking forward to 2001—when generation would no longer be regulated—the Alberta government also created power purchase arrangements (PPAs). The PPAs applied to existing power facilities: distribution companies would purchase electricity from power generation companies in a competitive market, and then sell the electricity directly to consumers.

TransAlta benefitted from the PPAs because they provided long-term and stable revenues for the life of the contracts with the distribution companies, which varied in length from three- to twenty-year terms. At the same time, the contracts allowed the company to earn extra revenue based on its productivity and on the sale of additional energy on the open market, and PPA holders could benefit from increases in the market. As well, investors liked the security of contracted sales that the PPAs afforded.

By 2003, it was apparent that TransAlta's PPAs in Alberta were an important part of the corporation's strategic plan. Overall, the company had long-term contracts for 85 percent of its power. Even though the price of natural gas was extremely volatile, and an increasing percentage of the company's generation was coming from gas, its contracts contained a clause that allowed TransAlta to pass on variable costs to the customers, and it had price caps on some of its natural gas contracts.

The stability of long-term contracts allowed TransAlta to make strategic decisions about decommissioning some of its older generating plants as part of the company's sustainable-development strategy. The company decided to retire its first unit at Lake Wabamun when its contract ended in 2003, and, in 2010, it retired its oldest thermal generating plants that dated back to the 1950s.

Long-term contracts were not limited to the Alberta market. In 2007, TransAlta extended its long-term contract with the Ontario Electricity Financial Corporation for the Ottawa Cogeneration power plant until 2012. The company signed a twenty-five-year contract with New Brunswick Power Distribution and Customer Service

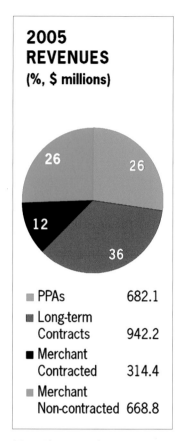

TransAlta Annual Report, 2005.
TRANSALTA COLLECTION

Corporation to provide wind power from TransAlta's Kent Hills project, starting in 2008. And, in 2009, the company negotiated a new contract with the Ontario Power Authority to extend the sale of electricity from the Sarnia power plant to 2025.

Selling on the Open Market

Energy marketing—selling electricity on the open market—had long been a key element in TransAlta's operations. As deregulation arrived in Alberta, the company's existing generation was grandfathered into the system and remained part of the current rate structure. But, it became evident that electricity from new generation facilities would be sold in a marketplace that was more competitive than in previous years.

"A priority is to build our capabilities in areas where deregulation is creating significant new opportunities, such as energy marketing," stated Senior Vice President of Sustainable Development, Jim Leslie. Selling electricity on the open market was already part of the company's operations in New Zealand in 1995. That same year, TransAlta began to pursue similar opportunities in Australia. And, in recognition of the growing importance of wholesale energy marketing, the company appointed Dawn Farrell as its first vice president of Marketing and Business Services.

"We have set aggressive goals to achieve growth targets for our non-regulated businesses," said Farrell at the time. "We plan to double the IPP megawatts we operate by 2001." In its first full year of operation, the energy marketing division was the sixth largest energy marketer in the U.S. Pacific Northwest, and the eighteenth largest in the U.S. It set an ambitious goal to provide 10 percent of the corporation's earnings by the year 2000. Though it more than met this target, its $54 million contribution to TransAlta's $187 million bottom line was largely due to volatility in the electricity and natural gas markets. The company emphasized that the long-term value of trading and marketing was that it could reduce its risks, find market opportunities—the Centralia operations were an early research success for this group—and enhance the value of the output from its generation facilities.

By 2000, TransAlta had energy-marketing offices in Calgary, in Portland, Oregon, and in Annapolis, Maryland—where, in 2000 and 2001, it purchased the Merchant Energy Group of the Americas (MEGA), a move that allowed it to build its presence south of the border. When amalgamated into TransAlta's operations, MEGA became part of TransAlta Energy Marketing (U.S.) Inc., a subsidiary of TransAlta Energy. In 2003, TransAlta closed the Annapolis office and moved its eastern trading operations to Calgary to reduce costs and to better monitor the risks. It also ended its participation in the New York Independent Systems Operator (ISO) auctions for transmission congestion contracts.

The energy marketing division brought in a net income of tens of millions of dollars every year from 2004 to 2009. In 2008—the year the price of oil hit an all-time high and the world economy boomed—it brought in almost $80 million. In 2009, the net was a relatively modest $44 million, due to the fall in the price of oil and natural gas, as well as the international economic downturn. Given that neither year's events were predictable, the ability of the marketing division to steadily contribute to the bottom line proved its importance.

In addition to the profits the energy marketing and trading division provided, marketing also played an increasingly strategic role throughout the decade. In 2005, TransAlta successfully negotiated contracts for 90 percent of its Centralia plant's 2006 output, and 85 percent of the 2007 production. At its Sarnia operations, it signed a five-year contract for 83 percent of its production with the Ontario Power Authority. The division's target of contracts for at least 75 percent of production was more than met.

By 2009, TransAlta had 95 percent of its power contracted, and in the face of weak economic markets, it still retained an 89 percent contract rate for 2010 and 83 percent for 2011. The corporation continued to manage its marketing wisely, entering into medium- to long-term contracts for most of its power. As of 2009, about 75 percent of its capacity was contracted for the next seven years, and its target was 90 percent in 2010. In the face of a turbulent economic climate, TransAlta's continued commitment to developing and maintaining its markets proved pivotal.

Centralia and the Move to Generation

Once deregulation took effect in Alberta in 2001, TransAlta had to market its electricity through PPAs. The company sold off its distribution and retail divisions in 2000 as part of its strategy to become leaner, trimmer, and more focused on its core strengths. In the new business climate, these segments simply could not meet the company's higher-return objectives. "We believe you don't have to own transmission assets to be a successful generator," the 2001 Annual Report noted, "but you do need to know the transmission business and its impact on electricity prices and markets." TransAlta's long history as a vertically integrated utility provided it with the wisdom to continue positioning itself well in the changing marketplace. The company also completed the sale of its New Zealand operations in 2000. The $1.46 billion realized from the sale of these divisions allowed the company to pursue its narrower strategic plan. The sale of the transmission assets in 2002 were in aid of the same goal.

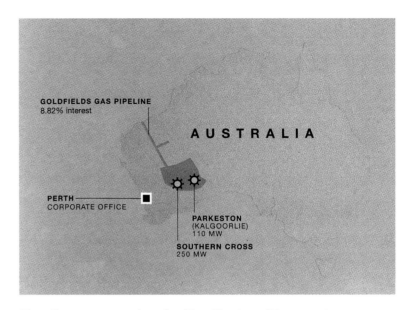

TransAlta operations in Australia, TransAlta Annual Report, 2001.
TRANSALTA COLLECTION

Energy Marketing Revenue Sources, TransAlta Annual Report, 2000.
TRANSALTA COLLECTION

THE TRANSFORMATION OF TRANSALTA

1997	2000	2001	2002	2003	2004
Integrated Regulated Utility	Sale of Distribution and Retail	Alberta Deregulation	Sale of Transmission	Generation and Wholesale Marketing Company	

The transformation of TransAlta, TransAlta Annual Report, 2004. TRANSALTA COLLECTION

continued on page 138

Centralia

"Few people knew the TransAlta coal mine outside Centralia was shutting before the gates actually closed Monday," a *Seattle Times* reporter wrote on 1 December 2006. "But a lot of people saw it coming."

Closing the thirty-five-year-old mine put almost six hundred people out of work, just before Christmas. TransAlta's mine and generating facility had been one of the largest employers in the county. Expert observers said it was inevitable that the state's last mine would close. The mine was old, the coal was poor, and costs were high. Finally, major cave-ins buried part of the coal operations.

TransAlta had worked hard to be a good corporate citizen since it bought the mine and electrical generating plant in 2000. In 2003, it gave $1 million to Centralia College to help fund job training. And in 2006, it gave the Lewis County Economic Development Council one thousand acres for an industrial land bank and spent $200,000 on site preparations. As part of its tradition of being involved with United Way programs where it operates, it contributed $500,000 to the charity in 2006. Between 2000 and 2006, TransAlta was the largest single donor in Centralia and had accounted for more than a third of the money raised by the United Way there each year.

After announcing the layoffs in late 2006, TransAlta also spent more than $5 million on job training programs and a community transition fund, as well as other assistance programs to help people transition into new jobs. Within a week of the announcement, the company set up a job fair where more than forty employers looking to place people in two thousand positions took resumes from the unemployed.

A year later, the sting was not yet completely gone for a region also hit by the loss of forestry jobs, but as the local newspaper, *The Chronicle*, noted, even with the layoffs in late 2006, TransAlta's coal-fired steam plant remained one of the largest employers in the county.

In the summer of 2007, TransAlta re-hired eighty of those laid off for reclamation work, and thirty-four other former employees found new permanent jobs with the company.

The Centralia College newsletter, *Blue and Gold*, commented on the mine closure in May 2007, noting that with government, college, and TransAlta help, one hundred students were able to enroll in classes. The company gave $500,000 to the college to assist family members of the dislocated miners to receive training.

The company continues to support the Centralia community, most recently with a $500,000 contribution to Centralia College. A new seventy-thousand-square-foot building, called the TransAlta College Commons, will bring together under one roof the admissions centre, student support services, computer labs, meeting rooms, and an expanded cafeteria. It is scheduled to open in 2013.

Mining crew at Centralia mine, 2008. TRANSALTA COLLECTION

With cash in hand, the company explored acquisition opportunities. Coal was what TransAlta knew best, and it wanted to generate electricity for the booming U.S. market. Part of its strategic plan was to add 500 MW of capacity each year in the Pacific Northwest region of the United States. Building on two years of market intelligence collected by the marketing division, TransAlta bought a project in May 2000.

The 1340 MW coal-fired Centralia generating plant and mine in Washington State represented an increase of 40 percent in coal-fired generation, and instantly gave TransAlta a stronger presence in the state of Washington. The company's generating experts went to work on its first coal-fired facility outside of Alberta and added millions of dollars in improvements. The result was an increase in capacity, productivity, and environmental upgrades. Centralia became one of the cleanest coal-fired plants in North America. By the end of 2000, the Washington plant was producing 17 percent of the company's power.

Electricity from this asset sold on the deregulated open market. But with Centralia came some obligations, including the liability to restore the mine site at the end of its life, as well as the need to install sulphur dioxide (SO_2) and nitrogen oxide (NO_x) scrubbers, which allowed the plant to burn more coal from its Centralia mine and less imported coal from the Powder River Basin in Wyoming. Applying the company's expertise, TransAlta increased coal production at Centralia by 35 percent in 2002, and by moving from a dragline mining process to a more appropriate truck-and-shovel method, it increased coal production and further reduced the cost of the fuel.

In 2006, TransAlta closed its Centralia coal mine, as the pits had reached the end of their economic life, and relied completely on coal from Wyoming. After these changes, Centralia became a stronger asset. The company improved its plant so that it could use higher quality coal, signed long-term sales contracts for the electricity, and prepared to invest in the technology necessary to capture 70 percent of the mercury produced during generation by 2010. It was also able to keep its operating costs below the rate of inflation. And in 2007, TransAlta voluntarily installed continuous emissions-monitoring systems at the plant.

The plant at Centralia produces 10 percent of Washington State's electricity, and, since 2010, TransAlta and the government of Washington have agreed to work together on a plan to deal with greenhouse gas emissions. A bill outlining the details of this plan was passed in April 2011.

A Portfolio Approach

In 2000, as part of the company's strategic initiative to diversify its fuel portfolio, TransAlta expanded its research into renewable energy and invested heavily in innovative projects. As part of its investments in non-regulated power generation, in 2000 the company created the $100-million Sustainable Development Research and Investment Fund. Its goal was to invest in carbon offsets and renewable energy projects and the development of new technology. This program examined 165 initiatives and invested in wind power and small, off-the-grid, distributed generation projects. Realizing that venture capital was not its business of choice, TransAlta eventually closed the fund. However, the experience proved useful as an exploration of alternative power generation.

One such initiative was the purchase of Vision Quest Windelectric of Calgary. Vision Quest then was expanded to include forty more wind turbines, and in 2001 TransAlta elected to use the electricity generated from this renewable source to power its head offices in downtown Calgary—a leading-edge move in power consumption.

Though the very small distributed generation projects were not a success, wind generation fit into the company's strategic plan in a number of ways. In 2002, less than 3 percent of the company's ca-

pacity came from its renewable energy sources. It set out to increase green generation to 10 percent of its portfolio by 2010, using wind as a short-term tool in that quest. Other projects would more than double this target within the decade.

In 2003, TransAlta invested $320 million to purchase a 50 percent interest in CE Generation LLC. CE Gen's gas-fired generation facilities in New York, Texas, and Arizona were nothing new for the Canadian utility, but the deal also included 10 geothermal units in California, in which naturally occurring steam produced deep under the earth's crust, drives a turbine. These units produced 253 MW of electrical generating capacity each year. Geothermal energy, along with wind energy, helped TransAlta reduce the emissions of its fleet and served as voluntary carbon offsets for its coal facilities.

During the first decade of the century, the company continued both to acquire and to build additional wind farms in Alberta, at McBride Lake and Summerview. The advantages were many, including quick construction and no fuel costs. However, wind farms are expensive to install and naturally can only be erected in locations with steady winds. By 2004, generation from renewables increased more than 77 percent. The renewable generation projects, along with improvements in plant performance in the rest of the operation, allowed the company to reduce its greenhouse gas intensity by 11 percent compared to 2000 levels.

The largest single jump in TransAlta's renewable capacity came in late 2009 with its acquisition of Canadian Hydro Developers. Though some speculated that TransAlta bought the company as a hedge against future carbon restrictions, Steve Snyder referred to TransAlta's long-time goal of diversifying its generating capacity and insisted that the electricity generated by Canadian Hydro's facilities made sense economically, regardless of government regulations or other developments. The acquisition of additional wind farms across Canada—in New Brunswick, Quebec, and Ontario—in the last half of the decade made the company the largest developer of wind power in Canada. As of April 2011, the generation capacity of TransAlta's wind portfolio was 1100 MW. TransAlta acquired the expertise to develop these projects at the lowest capital cost of any company in the country. It also expanded the geographic scope of its wind portfolio from three provinces to six.

Snyder believes his team is making an important contribution to the company's future by developing a three-pronged plan. The strategy is in place for the next twenty years, with alternative plans mapped out should they become necessary. First, TransAlta has to reduce its carbon footprint. It is increasing its focus on renewable energy sources and the cutting-edge clean coal technology of Keephills 3. With the addition of Canadian Hydro assets, TransAlta's renewable portfolio grew to 23 percent of its total energy portfolio. Second, it is being careful not to become locked into natural gas, because that carbon-based fuel—though cleaner than coal at this time—has, historically, proven expensive and volatile. And third, under Snyder's leadership, the company is keeping "the coal option open, because at the end of the day coal will still be—if we can get the right technology—the low-cost, reliable producer that uses our natural resources the best." If one of these three priorities changes, the company is in a good position to adjust because of its broadly based fuel supply.

TransAlta's balanced portfolio assures strength and options for the future. It offsets carbon costs; it will attract customers who want to contribute to society by purchasing cleaner power, even at a higher cost; and it will attract shareholders who desire a green investment. Providing the lowest-cost electricity was once the company's most important goal. In a broader and more nuanced culture on the part of both consumer and investor, TransAlta has learned to listen to the people who want to be associated with its reputation, not only as a reliable company, but also as a corporation that is progressive and leads by example.

Imperial Valley, California, geothermal operations, 2010. TRANSALTA COLLECTION

Sustainable Development

Calgary Power's presidents and directors would probably smile at the suggestion that it was only in the 1980s that TransAlta's leaders first began addressing sustainable development. After all, everything the earlier leaders did was sustainable development, intended to sustain the development of the utility company.

"Providing electrical service necessitates undertaking activities which impact the environment," wrote president Ken McCready in the 1988 Annual Report. But, for this new leader, from a new generation, meeting environmental responsibilities to counteract the impact was an additional long-term goal. And it was becoming clear that the environmental side effects of burning fossil fuels—natural gas and coal—in the company's thermal plants would draw critical attention from the public and from government at all levels. Given that Canada has virtually unlimited supplies of the natural resource, and that many of the industrialized and developing countries where TransAlta has diversified rely heavily on coal, TransAlta had to develop a plan to maximize the potential of the fuel and also meet environmental concerns. Technology was the key.

In 1988, TransAlta and a handful of other coal-burning electrical utilities, together with Alberta Environment, the Canadian Petroleum Association, and the Energy Resources Conservation Board, commissioned a study into the possible effects of the combustion of coal on Alberta's environment. Due to the low sulphur content in the province's relatively hard coal, the study concluded, "There is no evidence of negative impact on soils, plants, or water as a result of acid air pollutants in Alberta."

But coal-fired plants release more than just acid rain, and so, in the late 1980s, TransAlta began paying attention to greenhouse gases, including carbon dioxide—a product of the combustion of any fossil fuel. A 1989 mandate of the board's Risk Management Committee was to pay particular attention to the company's environmental safeguards. An Environmental Advisory Panel was set up, composed of Alberta residents with a broad range of backgrounds and skills, to help TransAlta develop policies and plans for its future.

Concern for the earth's health and the company's future finally came together explicitly in 1990. "TransAlta is committed to the environment and sustainable development," the company wrote in its first Environmental Policy Statement. "Protection of the environment is a vital element in our business." TransAlta's Environmental Policy Statements set goals for reporting on the environmental impact of its operations, for supporting socially responsible environmental standards, for encouraging conservation and efficiency initiatives and environmental education programs, for protecting the health of its employees and the public, and for seeking profitable businesses opportunities that would enable solutions to environmental problems.

In the early 1990s, the company continued its advocacy of sustainable development within TransAlta, within Alberta, and beyond. Jim Leslie headed the campaign, and TransAlta appointed Bob Westbury as its first vice president of Environmental and Public Affairs. After the 1992 Rio Earth Summit, leaders from around the world formed the World Business Council for Sustainable Development, and TransAlta President and CEO Ken McCready—one of the founders of the council—made sure the company was involved from the start. The Alberta government also named McCready chair of the Alberta Round Table on Environment and Economy. That same year, the company announced a $500,000 grant to create a sustainable development program in the Faculty of Environmental Design at the University of Calgary.

On 1 September 1993, the Alberta Environmental Protection and Enhancement Act came into effect. For TransAlta, the government act required increased levels of public participation in Environmental Impact Assessments conducted on its proposed projects. But the corporation was confident that its environmental policies were capable of satisfying the environmental concerns and expectations of customers, investors, and the public.

Finding a cost-effective way to address greenhouse gases was becoming a major challenge. In April 1994, TransAlta made a commitment to stabilize its net intensity contributions of these gases at the 1990 level by the year 2000. This goal was met in 1998. Leslie noted that the company supported market-based solutions to the global-warming problem. Command and control prescriptions by governments merely lead to a host of detailed instructions, but setting goals and encouraging companies to meet them by employing the best technology and innovative ideas benefits everyone.

TransAlta's board has always been charged with overseeing

continued on page 144

TransAlta's People

Gone are the days when the little Alberta utility company had a small office in a Quonset hut down by the Centre Street Bridge in Calgary's Chinatown district, while a few men lived and worked at its generating dams along the Bow River, out west toward the Rocky Mountains. In 1940, the company had 160 employees, its head office was in Montreal, and wartime restrictions meant the company's vehicles ran on bald tires—when they could get gasoline.

Today's TransAlta boasts a workforce of more than 2,200 people in three countries: Canada, the United States, and Australia. In an effort to help employees better understand the direction of the company, CEO Steve Snyder started hosting employee town hall meetings in 2009, and COO Dawn Farrell hosts employee sessions each fall, similar to the coffee pot sessions of twenty-five years ago.

Human Resources: Keeping the corporate culture alive, evolving, and engaged while meeting the challenges of changing times is tricky. Over the years, TransAlta has employed a variety of initiatives to foster a sense of community. In spite of the change from a corporate culture focused on regulation to an entrepreneurial culture, some of the company's longstanding relationships with its workers endure.

Unionized workers, represented by national and international unions and performing skilled tasks in the company's power plants and coal mines, account for more than half the labour force at TransAlta. TransAlta maintains good relations with its unionized employees, despite challenging market conditions, seeing its unionized employees and their unions as business partners. A gain-sharing program of bonuses and incentives is in place for these workers. The company has only once been subjected to a strike, for a month in 2007 at a single power station.

Non-union employees also benefit from incentive pay when annual goals are achieved. Over the years, the company has introduced numerous employee recognition awards, the most recent being the President's Awards for Safety. Stock option plans and share-purchase plans are in place, and the pension plan is strong, as is the company's education-assistance program, which supplies grants and scholarships for children of TransAlta employees. In 2006, TransAlta instituted a sabbatical program, allowing employees to take up to six months away from work every four years, at half pay and full benefits, to pursue other passions. TransAlta's employee benefits program features more than twenty-five benefit package options, so that employees can tailor benefits to their personal needs. Enhanced flexibility will also be part of the new benefits program to be introduced in 2011.

Employees as Volunteers: Community engagement has long been part of TransAlta's corporate culture, specifically through two programs engaging volunteers, known as TACT and POWER.

The TACT program—TransAlta Community Transformers—encourages employees to help their local communities through meaningful grassroots initiatives. Formed in 1999, the first TACT teams at the Alberta power plants were followed by teams at the Centralia plant in 2001, the Fort McMurray plant in 2002, and at Sarnia in 2004. TACT has supported a food drive, provided clothing for the poor children, and adapted canoes for the disabled, among other initiatives. The company's community-investment initiatives in Australia during 2007 helped fund the Royal Flying Doctor Service and sponsored a drive (involving a moustache competition) to raise funds to combat male depression and prostate cancer. In Mexico, it supplied emergency water and clothing to people affected by flooding and helped create a United Way program in the remote area around the company's Campeche plant. When Haiti was devastated by a hurricane in early 2010, TransAlta made an initial donation of $25,000 to the Canadian Red Cross Haiti Earthquake Fund and then matched contributions by employees for an additional $35,000.

Projects Organized With Energetic Retirees (POWER) involves a dedicated group of company alumni that volunteer their time. POWER logged five thousand hours of volunteer time in 1999 alone. That year, POWER volunteers delivered fourteen thousand pounds of fresh vegetables from its community garden, developed along an old Altalink transmission line, to people in need. Other years, they

TACT Volunteer, 2005. TRANSALTA COLLECTION

In 2005, TransAlta donated more than $3.6 million to community investment programs

have built desks at Calgary's Hull Child and Family Services, carried out highway cleanups, conducted historical walking tours in Calgary, repaired bicycles and buildings at the Children's Ranch in Kananaskis Country, knitted clothing items for premature babies as part of the Knits Wits program, and painted houses for elderly Calgarians as part of Calgary's Paint the Town program. "Getting physical exercise, being with friends, and doing something that is very worthwhile makes you feel good," volunteer Sandra Gilbertson recalled in 2003. "It's a way of giving back to the community."

Aboriginal Relations: Aboriginal relations have been important for the company since its earliest days, given that its first facilities were on land traditionally used by the Stoney people along the Bow River. In 1995, TransAlta created Aboriginal advisory councils with the Paul and Siksika bands and met their leaders to work together on collaborative projects. The company also hired its first Aboriginal Relations specialist in this time period. George Blondeau was instrumental in creating many of the programs in place at TransAlta today.

TransAlta helped sponsor Coyote Kids with the Bent Arrow Traditional Healing Society, an Edmonton-based program designed to empower Aboriginal children by providing them with traditional teachings, values, and beliefs. Between its inception in 1997 and 2003, the program served more than eight hundred children.

In 1998, TransAlta made a donation to the Eagle Spirit Development Corporation of Alberta for a centre to encourage Aboriginal people to develop new businesses. Scholarships for First Nations people were also made available that year, and 425 employees attended Aboriginal Awareness Workshops set up by the company. In 1999, TransAlta made its efforts to connect with First Peoples more intentional and sought ways to increase business dealings with them. As part of a concerted effort to hire Aboriginal people, it hired thirteen full-time workers in 1999 and another twenty-one seasonal employees as well as creating apprenticeship programs.

In 2002, TransAlta settled a land claim with the Stoney (Nakoda) First Nation and returned 114 hectares of land and mineral rights near the Horseshoe Dam on the Bow River west of Calgary to the community. The agreement also included a $600,000 development fund for the community's education and cultural activities.

In 2004, the company helped the Paul Band, west of Edmonton, conduct a traditional land-use study, along with government and industry partners. The two-year project documented traditional land use in order to preserve its cultural history and to assist in local economic and resource development. The band's oral history, passed down by Elders, was documented through tape-recorded interviews. Global positioning systems and other information technology helped capture data. Today, a traditional land-use study is conducted with local Aboriginal people as a part of every development project.

In 2006, TransAlta set up the TransAlta Advisory Committee with representatives from all the bands affected by the company's operations, and developed best practices guidelines for transmission systems through Aboriginal lands. And in 2007, it once again hired a full-time Aboriginal relations specialist, Kirby Smith.

In 2009, the company implemented a new Aboriginal relations policy to provide guidelines for relationship building and communications, and to standardize practices for all locations. Most recently, TransAlta worked with the Piikani Traditional Use Office to involve Elders in reviewing the work associated with the new Ardenville wind farm in southern Alberta.

the direction of the company and ensuring its health and growth. Anticipating issues and risks, encouraging management, and making complex decisions is a tricky mandate. Donna Soble Kaufman, a board member since 1989, recalled:

> I well remember my initial reaction when the notion of sustainability, as it relates to environment issues, was introduced to the TransAlta board in the early nineties. What will this mean? As directors of an energy company, we were intrigued by the notion of sustainability because we knew environmental issues would become increasingly important.
>
> We believed if TransAlta took proactive steps then, it would benefit not only our shareholders and other stakeholders, but the environment as well. This was something of a leap of faith, but we believed it was the right thing to do, and that conviction has proven correct many years later.

In the mid-1990s, the company applied its operational skills to the coal challenge. Improvement in the quality of coal allowed it to burn at a higher temperature, reducing its emissions as a result. TransAlta became even more active in the international arena and participated in the discussions leading up to the global warming conferences in Kyoto in late 1997 and in Buenos Aires in 1998. On the home front, it continued to be involved in discussions with federal and provincial agencies as they worked on policy issues and priorities.

In 1998, TransAlta was one of four companies worldwide selected by a Swiss think-tank to act as case studies in sustainable development for business schools at European and North American universities. The World Business Council for Sustainable Development promoted these case studies at the climate summit in Buenos Aires in 1998. The company also released its first annual Sustainable Development Progress Report.

Education is an important part of environmental stewardship, and so in 1999 TransAlta announced Project Planet, a challenge to Alberta's youth to take up the cause of sustainable development. The company recruited hockey legend Wayne Gretzky, who helped launch the program with students from St. Monica's School in Calgary. Project Planet, which ran for five years, attracted innovative environmental project submissions from hundreds of students, engaging them in sustainable development pursuits. Gretzky recalled that his father, Walter, coached him to "skate to where the puck is going to be, not to where it has been." The same advice applies to sustainable development initiatives, because they are not locked in time and place. For an individual, a company, or a society to be committed to sustainable development we must all look to "where the puck will be."

In 1999, TransAlta was named global leader on sustainability reporting for the electricity sector by the Dow Jones Sustainability Group Index (DJSGI). TransAlta was one of eighteen companies honoured by the index, which noted, "These sustainability companies pursue opportunities in a proactive, cost-effective and responsible manner today, so they will outpace their competitors and be tomorrow's winners." The DJSGI was the first global sustainability benchmark.

In January 2004, the World Economic Forum named TransAlta one of the top one hundred companies in the world—the only company in its sector in Canada— for its sustainability efforts. In 2006, the DJSGI began selecting the top 20 percent of companies in each sector out of the six hundred largest North American companies. TransAlta has been on that list every year. In 2010, TransAlta was the only Canadian company in the utilities sector to make the grade, and one of only twenty-four Canadian companies on the index.

TransAlta participated in the world's first-ever internet-based trade of emission-reduction credits in 1999 with a deal to purchase up to 2.8 million metric tonnes of carbon dioxide emission reduction credits from farms in the U.S. And in 2000, it conducted the first transatlantic trade of carbon dioxide emissions reductions with the German electric company Hamburgische Electricitäts-Werke SG. This 24,000-tonne emissions reduction trade was equal to the annual emissions from three thousand cars.

In the absence of regulatory action, TransAlta embarked on a voluntary program of reducing its CO_2 emissions. In its report to Canada's Voluntary Challenge and Registry (VCR) Program in 1999, the company noted that it had reduced net emissions per megawatt-hour of electricity generated to 812 kilograms from 827 kilograms the prior year and from 994 kilograms in 1990. TransAlta received the VCR's Champion Gold Level Award for its corporate action. As well, the company was one of the founders of the International Emissions Trading Association.

In 2001, as a founding member of the Canadian Clean Power Coalition (CCPC), the company set out to apply new technology to make coal-fired generation more sustainable. The CCPC set an ambitious goal: to build and operate a full-scale demonstration project for the removal of greenhouse gases and all other emissions of concern from an existing coal-fired plant by 2007. Though the coalition's goal of a zero-emissions plant was not met by the deadline, its studies into reducing emissions from coal plants showed that gas outputs from burning coal can approach the low levels of natural gas combustion. The project indicated that the potential for CO_2 storage might exist in the Western Canadian Sedimentary Basin. More research and development is needed to deliver on the promise of clean coal technology and low-cost gasification.

Carbon Capture and Storage: Project Pioneer

In 2008, TransAlta proposed to install post-combustion retrofit technology at the Keephills 3 power plant already under construction. The plan was to create a large-scale demonstration plant to capture the CO_2 from the flue gas emissions stream. Capturing one million tonnes of CO_2 per year, the project will be the largest carbon capture and storage project (CCS) of this type in the world. The project could prove that existing coal-fired plants can reduce their emissions to levels lower than facilities that burn natural gas. This CCS included transportation and permanent storage in a mature oilfield to the west. Capital Power Corporation of Edmonton was a partner in the project, as it was a partner in Keephills 3.

The Alberta government has set a goal of capturing and storing five million tonnes of CO_2 annually by 2015. The government created a $2 billion fund to help companies apply CCS technology, still in development, at the major project level. The federal government similarly created a $1 billion Clean Energy Fund. In October 2009, TransAlta received more than $770 million from these funding initiatives and was able to proceed with the unique, large-scale "Project Pioneer" as a public-private partnership.

A Challenge to the Plan: The Luminus Group

Not everyone supported TransAlta's sustainable-development strategy. In late 2007, the corporation's single largest shareholding group tried to change the company's focus. To that end, Luminus Group attempted to install four of its own people on TransAlta's board of directors in order to force the corporation's hand. Luminous wanted TransAlta to sell off its renewable power generation and its plants in Mexico, and take on a large amount of debt in order to buy back its own stock and pursue conventional power-generation options.

In the face of this aggressive challenge to the company—a fact of life in tight capital markets—the board of directors held firm. The company continued its commitment to a low-to-moderate risk strategy and a strong balance sheet, with disciplined capital allocation aimed at growth, a consistent dividend and share buybacks. As a result, TransAlta continued its tradition of providing reliable and sustainable power to its customers and good value to its shareholders.

Keephills 3 under construction, 2010. TRANSALTA COLLECTION

In 2010, the federal environment minister announced the phasing out of coal-fired power plants. The new policy required that all coal-fired plants be closed by the end of their forty-five year life or meet the emission standards of a natural gas-fired facility. In the face of this new reality, TransAlta is building on its commitment to sustainability with improvements to its existing coal-fired facilities and installation of innovative technology in the plants that are currently under construction. The company is also drawing upon its diverse portfolio of fuels, expanding into more gas-fired generation, and generating an increased amount of power from renewable sources.

"TransAlta has a very clear strategy," Dawn Farrell said in 2010:

We have clarified the strategy in terms of growing the business. We got it focused on the west, and got it focused on renewables. As a result, we built more wind farms and bought Canadian Hydro, and really put in place some of the last legs of the strategy that we'd actually started in 1990, when we began thinking about carbon and doing carbon offsets.

We also continue to reposition the coal part of the business around the carbon issue. We've got the CCS project in place now, we've got almost 25 percent renewables, and we will continue to reposition our coal assets as we go forward.

Horseshoe Dam, 1911 TRANSALTA COLLECTION

Wolfe Island wind farm, 2011. TRANSALTA COLLECTION

A Part of the Solution

In 1911, the men who ran the hydro dam at Horseshoe Falls on the Bow River, feeding electricity to the nearby cement plant and to the City of Calgary, operated as a private corporation, with little direct control by government. Calgary Power integrated the best hydro technology, superior transmission control, and market certainty and was fortunate enough to have access to low-cost capital.

One hundred years later, much has changed, yet many of the same challenges remain. Most electricity in Canada is produced and distributed by large, government-owned utilities, though investor-owned utilities in Alberta, Ontario, and Nova Scotia compete in a more open market. Regulated operators now carry out transmission, for the most part, and in the face of growing environmental concerns, governments at the provincial and federal levels regulate the operations of all electrical utilities. Though technological improvements have reduced the intensity of emissions from the generation fleet, TransAlta and other Canadian utilities will need to make continued advancements in this area. TransAlta has been investing in carbon-offset credits, technology funds, and renewable power as authorized under the Alberta regulators. And the company is pursuing renewable energy alternatives in order to stay ahead of the ever-changing regulations and to serve the public interest. It continues to work with governments to address issues proactively, helping to encourage the creation of regulations that allow companies to apply their ingenuity to the challenges facing society and industry.

Hard-nosed pragmatists that they were a century ago, Max Aitken and Izaac Walton Killam also believed that at the core of every business was a commitment to society. Although they might hardly recognize the company's modern geothermal operations and windmills, the men who, a century ago, built the first Calgary Power hydro plants were entrepreneurs. Providing electricity to a grid that was only then coming into being, they were committed to the future. It was they who created the strong foundation for TransAlta to move to renewable sources of power and sustainable development. Successors Geoff Gaherty, Marshall Williams, and Ken McCready guided the company through periods of unprecedented expansion and on-going challenges. More recently, Steve Snyder has moved the investor-owned utility from a provincial presence to an international player; from a coal-based producer of electricity to one with a diversified portfolio that includes coal, gas, and renewables; from a regulated utility to a mix of long-term contracts and merchant plants that capitalize on market opportunities. TransAlta has become a world leader in the development of energy technologies that produce power responsibly.

Steve Snyder is as much a visionary as the men who committed the company to hydropower in the early 1900s, and to coal in the 1950s. "We still ultimately believe we need and can find a way to use that carbon resource," he said of coal. "The industry is about to go through its biggest transformation—even bigger than deregulation. A whole technological revolution is on the way." Snyder's trust in technology to reduce greenhouses gases is solidly rooted in scientific research and practical experience.

"It's not the fuel that is the problem—it's the emissions. It is clear that we need a broad portfolio of fuels to meet our nation's future energy demands … and, honestly, that includes hydrocarbons," Snyder said in late 2010. "The world relies on fossil fuels and will continue to rely on fossil fuels for the rest of our lives and likely our children's lives. But I say this with a caveat. There's an expectation that, as a country, we should move toward an increasingly lower carbon future. And I don't dispute that. The biggest challenge for Canada's energy sector is figuring out how it can grow the industry while, at the same time, reducing emissions and other environmental impacts."

This is a significant challenge, but the history of TransAlta is one of challenges faced and challenges overcome. Innovation is clearly necessary to solve the tough problems of the future, as the global demand for energy reaches unprecedented levels and the nearly two billion people around the world who have little or no access to electricity aspire to a better quality of life. TransAlta sees itself as part of the solution. As in the past, the company's well-trained, experienced, and dedicated people will guide it through the periods of change and evolution that lie ahead and deliver on Steve Snyder's vision. The little utility that started out making electricity with one hydropower generator in 1911 has always been willing and prepared to succeed when confronted with historic challenges.

"Transitions of this magnitude take time," says Snyder. "But, despite what the naysayers claim, I believe we have the time to get it right.

"What we don't have time for, is to get it wrong."

Afterword

The Way Forward: More Electricity, Fewer Emissions

Historians predict the past, and then only with great care. But CEOs are called upon to lead their corporations into the future. As the leader of TransAlta, it is my job to work with the board to take the company forward in a manner that serves the best interests of our customers, employees, and investors. Working with an energetic team of innovative people, building upon our legacy of valuable assets, and taking the time to make the right decisions, I am confident that TransAlta is ready to meet the challenges of the next century.

For one hundred years, TransAlta and its people have been building a tradition of operational excellence, responsibility, and commitment to community. That legacy will allow us to continue to earn the right to operate our plants in communities, serve our customers' needs, and deliver value to our shareholders for the very long term. We will continue growing the company by developing our existing markets, diversifying our assets and fuel sources, maintaining a strong balance sheet, paying a strong dividend, and always planning into the future.

Making the best use of our valuable assets includes managing our existing facilities wisely, and running these assets for as long as economically profitable. Our people work hard to ensure that our plants operate as efficiently as possible, and though we already have one of our industry's lowest injury-frequency rates, we continue to strive for a target of zero accidents in the workplace. We have demonstrated that we are environmental leaders in our industry, and we will continue to take pride in our initiatives to increase the energy efficiency of our operations, reduce harmful air emissions, and invest in the best technology.

TransAlta's first electricity came from hydro plants along the Bow River in southern Alberta more than one hundred years ago. As demand for electricity increased and as opportunities for more hydro development in Alberta dried up, we embraced an abundant and local fuel source to help us continue to power Alberta's growth and economic development. Ever since, coal has been a central fuel in our portfolio, and we remain committed to finding ways to utilize this relatively inexpensive and abundant source of energy in a responsible manner, preserving its energy potential, but minimizing its harmful emissions.

To that end, we were the first to build super-critical coal-fired facilities in Canada, and our Centralia facility is one of the cleanest coal plants in the United States. All industrial emissions in our society need to be managed and reduced, and we continue to demonstrate corporate leadership by pursuing creative solutions to coal's challenges. Among other things, we are working with governments and global technology leaders to develop and commercialize carbon capture and storage. Many challenges remain in this area, but we believe that hard work and innovation can create an environment where coal will be a cleaner fuel.

At the same time, TransAlta is Canada's largest producer of renewable power. More than half of our production today comes from renewable sources and from cleaner-burning natural gas, with more than 1957 MW of wind and hydro power across our fleet. These renewable energy initiatives include century-old hydroelectric plants on southern Alberta rivers as well as new wind farms across Canada. We were the first Canadian energy company to participate in a wind power project on a major scale, and we are currently one of Canada's largest wind energy producers. Since 2000, we have invested nearly $1 billion in other forms of renewable energy, including geothermal facilities. We intend to continue to build TransAlta's portfolio of renewable fuel sources.

Looking ahead, I see great opportunity for our company and shareholders, and challenges that we must work hard to address. In a dynamic industry and operating context, our company will need to continue to grow in order to succeed. Electricity will remain in high demand, and we will need to pay close attention to our business—both the assets in our existing fleet, and potential new opportunities—to ensure that we continue to serve our customers. As this book demonstrates, our company has always looked ahead, and we have grown with the times. At our centennial, I feel that we must continue to do so. We will grow in a manner consistent with our tradition and with the values that underpin our company. TransAlta will grow, but it will do so prudently, respecting our shareholders, maintaining the strength of our balance sheet and dividend, valuing our employees, and becoming part of the communities in which we operate.

As we move forward, regulation of emissions will continue to be an important element of our decision-making. While this can introduce some measure of uncertainty, we feel that clear rules and a price on carbon will help our company continue to meet society's demands for energy in the most responsible way possible.

Our history and strategy position us well for the future. The expertise that our employees possess in renewable power, in natural gas, and in cleaner-burning coal presents our company with an array of paths to follow. Power generation is a capital-intensive industry that requires long-term thinking. But our company's diversity gives us options, and makes us nimble. We begin our next century in a very strong position.

When I look at our company today, I see a talented, committed, and excited group of people. From our headquarters in Calgary, to our wind operations in Quebec and New Brunswick, to the United States and Australia, we have a team that is poised to capture the opportunities of the future. We are determined to build on the legacy presented in this book, and to go on powering generations and helping communities prosper for the next one hundred years and beyond.

Steve Snyder
Chief Executive Officer
2011
Calgary, Alberta

Index

A

Abelseth, Jal, 55
Aboriginal relations, 35–6, 51, 52, 143
Abraham Lake, AB, 98
advertising, 61–3, 85, 88–9, 105. *see also* marketing
Aitken, Max, 4, 25, 27–30, 31, 32, 39
Alberta and Great Waterways Railway, 24, 26
Alberta Power Commission, 66–7, 69, 98
Alexander, W. M., 28
Armistead, Bob, 30
ATCO Corporation, 107, 110

B

Barrier Lake Dam, 55, 77
Bearspaw, David, 36
Bearspaw, Johnny, 36
Bearspaw Dam, 36, 80
Beil, Charles A., 63
Bell, Alexander Graham, 11
Bennett, R. B.
 biography, 25
 with Calgary Power, 28, 29, 30, 31, 32, 35, 36, 39
 as lawyer, 21, 24
 as politician, 24, 26, 45
Bighorn Dam, 98
Biles, Herb, 54, 58
Blondeau, George, 143
Bochek, Linda, 90
Boissoneault, Ray, 45
bombing of transmission lines, 110–11
Bow River, 26–8, 80
Bradley, Hank, 78
Brazeau Dam, 95, 98
Bridge, Will, 109
Brown, Fred, 70
Brownie, Bob, 45
Buchan, Fred, 56
Budd, W. J., 28
Burns, Pat, 22

C

Calgary, 12–13, 16–17, 19–22
Calgary Electric Lighting Company, 17–18, 19
Calgary Power
 beginnings, 32
 change of name to TransAlta, 104–5
 and international markets, 104
 market expansion outside Calgary, 42, 52, 54
 move of head office, 74, 83
 public trading of shares, 85. *see also* TransAlta Utilities Corporation
Calgary Water Power Company, 18, 22, 42
Cameron, Hugh, 78
Camrose-Ryley, 101
Canada Cement Company, 28, 54
carbon capture, 145, 146
carbon offsets, 129, 138–9, 146
Cascade power plant, 65, 77
Centralia, WA, 135, 137, 138
Chace, W. G., 28, 30
coal as power source
 mining process, 96–7
 move to, 93–5
 new technology in, 139, 144, 145
 in US, 137–8
cogeneration, 119, 122, 130–1, 134
Commission of Conservation, 26–7
community outreach, 51, 85, 90, 132–3, 137, 142–3
corporate governance, 125–6
Cross, A. E., 38
Cross, J. B., 38
Cundall, John, 79

D

Davies, Donald, 17
Dean, Audrey, 62
decommissioning plants, 134
Depression, 45–6, 52–3, 56, 57, 60
deregulation, 124–5, 126, 129, 130–1, 134–6
Dowling, Don, 78
downsizing, 128, 134, 137
Drury, Victor, 40

E

Edison, Thomas, 5–6, 7, 9, 11
Elbow Falls, AB, 35
Electric Energy Marketing Agency (EEMA), 119, 121, 130
electricity, discovery and development of, 3–11
employee programs, 51, 60, 120, 142
engineers, hiring of, 56, 95
environmental stewardship
 commitment to sustainability, 113–14, 121, 133, 138–41, 144–6, 148
 early efforts in, 59, 96
 regulatory assessments of, 100, 108–9
 and renewable energy, 2, 114, 116–18, 135, 138–9, 147

F

Farm Electric Services, 69, 72
Farrell, Dawn, 134, 135, 142, 146
Finnis, Tim, 96
fires, 56–7
fly ash, 114

G

Gaherty, Geoffrey
 biography, 50–1
 with Killam, 40, 42
 as president, 52–3, 56, 63, 68, 74–5, 93
Gale, Fred T., 59, 67
Garner, George, 85
General Electric, 5, 8, 9, 11, 39
geothermal energy, 139, 140
Ghost Dam, 1, 30, 42, 43, 80
Gilbertson, Sandra, 143
grain elevator electrification, 56
Gretzky, Wayne, 144

H

Halpen, Mike, 87
Hamilton, Jacques, 91
Hansen, Darrel "Hap," 47, 50, 68
Haskayne, Dick, 130
Hay, William, 78
Holt, Herbert, 31–2, 33
Horseshoe Falls, 28, 30–2, 35, 57, 147
hostile takeovers, 107, 110, 145
Howard, A. W. "Bert," 84, 85, 86, 105, 118
Howe, C. D., 64
Hubbard, Elbert, 5
hydro as power source
 on Bow River, 26–8
 early efforts, 9, 11, 18, 19, 20–1, 22. *see also* specific dams and power plants

I

independent power plants (IPPs), 122, 130–1, 134
Interlakes power plant, 80
International Brotherhood of Electrical Workers (IBEW), 87

K

Kananaskis Falls power plant, 35–6, 37, 52, 53, 65
Kaufman, Donna Soble, 144
Kawalchuk, Mike, 59
Keating, John, 117
Keating, Ross, 117
Keephills, AB, 101, 108–9, 110
Keephills 3 power plant, 145, 146
Kent Hills, 135
Killam, Izaak W.
 biography, 48–9
 as chairman, 2, 46, 52
 death, 83
 with Gaherty, 40, 42
 and Royal Securities, 39
King, George, 17, 19

L

Lake Minnewanka, 36–7, 42, 64–5
land reclamation, 114, 115
Laskoski, Ray, 85
Lebourveau, Homer, 92
Leduc No. 1, 72, 73
Leslie, Jim, 135, 141

logos, 79, 90, 102, 105, 123
Lougheed, James, 21, 24
Lougheed, Peter, 99–100
Lyall, Scotty, 41

M

Macphail, Andrew, 35
Malraison, R., 53
Manning, Ernest, 66–7, 68–9, 99
marketing, 42, 52, 54, 135–6. *see also* TransAlta Utilities Corporation
Martin, E. Bruce, 67
McColl, Gordon, 85
McCready, Ken
    biography, 121
    joins Calgary Power, 86
    as president, 59, 104, 120, 126, 140, 141
McGowan, Bob, 59
McKenzie, Gordon, 68, 72, 76
McLeod, Ernie, 65
Millar, Ken, 65
Milligan, G. H., 54, 92
Modern All-Electric Kitchen display, 62–3
Montreal Engineering, 30, 36, 48, 49, 85
Moore, A. W., 45
Moore, Dolly, 54

N

National Energy Program (NEP), 104
Nu-West Development Corporation, 107
nuclear power, 118

O

oil field development, 61, 72, 73, 76
Ontario Hydro, 20–1
Our Alberta Heritage, 91

P

Parkinson, George, 71
Pearson, Marianne, 62, 63
plebiscite on public ownership, 68–9
Pocaterra, George, 53
Pocaterra power plant, 80, 81
Polutnik, Tony, 59
Power Purchase Arrangements (PPAs), 130, 134–5
Prince, Peter, 18, 19, 21, 22
Progressive Conservative government, 99–100, 101, 121
public consultations, 108–9
Public Utilities Board (PUB), 39–40, 107

R

rate increases, 100
Rau, Verlin, 90
recession, 103, 104
Reeves, Bill, 59
regulatory assessments process, 100, 101, 108–9, 110, 118–19, 141
The Relay, 58–9
renewable energy, 2, 114, 116–18, 135, 138–9, 147
Robertson, Charles, 54
Robertson, F. J., 54
Robley Jr., Ted, 66
Roebotham, A. E., 41
Roper, Elmer, 68
Rowland, Gerald, 70
Royal Securities, 52, 83, 85, 102
Rundle Dam, 80
rural electrification, 60, 66–73, 77
Rutherford, Alexander, 24, 26

S

safety, 78–9, 133
Sandilands, Dick, 85
Saponja, Walter, 86, 87, 95, 125, 126, 128, 130
Schaefer, Harry
    and B. Howard, 86
    biography, 127
    as CFO, 118
    as chairman, 125, 126
    and computers, 121
    on Gaherty, 51
    on Killam, 49
    and TransAlta Resources, 119
Schaefer, Rob, 127
Scott, Shirley, 62, 63
Seebe, AB, 54–5
Sexton, Jack, 50, 51
Sheerness power plant, 107, 110
Sherman, Harry, 51
Sifton, Arthur, 26, 39
Sikora, Lou, 58
Slater, Vicki, 59
Smith, Bob, 80
Smith, C. B., 28, 30
Smith, Kirby, 143
Snyder, Steve, 129, 130, 139, 142, 148, 149–50
Social Credit government, 60–1, 66–7, 69, 99
Spray Lakes power plant, 42, 64, 75, 80
stock market crash of 1929, 45
Stoney (Nakoda) First Nation, 35–6, 51, 52, 143
Strappazzon, John, 36, 45
Sundance Plant, 98–9, 114, 115
sustainability. *see* environmental stewardship

T

Tapics, John, 86, 114
tar sands, 99–100
Taylor, Harold, 87
Tesla, Nikola, 6
Thomas, Stu, 78
Thompson, G. H. "Harry," 58, 85, 91, 92
Thompson, Scotty, 70
Three Sisters Dam, 80
TransAlta Energy Systems, 120, 130–1, 135
TransAlta Resources, 119
TransAlta Utilities Corporation
    created, 105
    diversification, 119–20, 122, 130–1, 134
    future of, 148–50
    headquarters, 105
    and international markets, 112, 122, 131, 135, 136–8
    interprovincial markets, 111–12, 134–5. *see also* Calgary Power
transmission lines, 110–13
Turner Valley oilfield, 61

U

unions, 87, 142

V

Victoria Park, 34, 36, 42

W

Wabamun Plant, 51, 92, 93–6, 98, 106, 134
Walker, James, 17
Walking Buffalo, 36
Westbury, Bob, 141
Westinghouse, George, 5, 7, 9
Williams, Flora, 62, 63
Williams, Marshall
    biography, 106
    and I. Killam, 2, 49
    joins Calgary Power, 86
    as junior executive, 93, 95
    as president, 103, 104, 105, 113, 120
wind power, 114, 116–18, 135, 138–9, 147
Wolley-Dod, Bill, 41, 58
World War I, 39
World War II, 57, 59, 63–4, 65–6
Wright, John J., 10